12TH
INTERNATIONAL
ANAMED ANNUAL
SYMPOSIUM

Spatial Webs: Mapping Anatolian Pasts for Research and the Public

© Koç University Press, 2021

Koç University Research Center for Anatolian Civilizations (ANAMED) Series

KOÇ UNIVERSITY SUNA KIRAÇ LIBRARY CATALOGING-IN-PUBLICATION DATA

Koç University Research Center for Anatolian Civilizations (ANAMED) Series

International ANAMED Annual Symposium (12th : 2017 : Istanbul, Turkey)

12th International ANAMED Annual Symposium: Spatial Webs: Mapping Anatolian Pasts for Research and the Public, November 29th-30th 2017 / Authors Ayşe Nur Akdal, Elton Barker, Anuja Dangol, Nicholas de Lange, Claudia Engel, Piet Gerrits, H. Murat Güvenç, Piraye Hacıgüzeller, M. Erdem Kabadayı, Vahakn Keshishian, Daniel Knitter, Turgay Koçak, Bernhard Ludwig, Dominik Lukas, Camilla Mazzucato, Osman Özkan, Jeroen Poblome, Michael Połczyński, Christopher H. Roosevelt, Amy Singer ; copy editors Lauren Davis, Jacob Chizzo ; publishing coordinator Alican Kutlay ; publishing assistant Alara Demirel ; book design Burak Şuşut ; symposium coordinator Şeyda Çetin; contributing coordinator Ebru Esra Satıcı ; project manager Buket Coşkuner ; ANAMED director Chris Roosevelt. -First edition- Istanbul : Koç University Press, 2021.

224 pages ; 19,5 × 25 cm.-- Koç University Press

ISBN 978-605-7685-37-7

1. Cartography. 2. Cartography--Turkey--History. I. Akdal, Ayşe Nur. II. Barker, Elton. III. Dangol, Anuja. IV. Lange, Nicholas de. V. Engel, Claudia. VI. Gerrits, Piet. VII. Güvenç, H. Murat. VIII. Hacıgüzeller, Piraye. IX. Kabadayı, M. Erdem. X. Keshishian, Vahakn. XI. Knitter, Daniel. XII. Koçak, Turgay. XIII. Ludwig, Bernhard. XIV. Lukas, Dominik. XV. Mazzucato, Camilla. XVI. Özkan, Osman. XVII. Poblome, Jeroen. XVIII. Połczyński, Michael. XIX. Roosevelt, Christopher H. XX. Singer, Amy. XXI. Roosevelt, Christopher. XXII. Çamurcuoğlu, Duygu. XXIII. Özgüner, N. Pınar. XXIV. Kutlay, Alican. XXV. Demirel, Alara. XXVI. Özgüner, Pınar. XXVII. Şuşut, Burak. XXVIII. Çetin, Şeyda. XXIX. Satıcı, Ebru Esra. XXX. Coşkuner, Buket. XXXI. Roosevelt, Christopher. XXXII.Title.

GA1303.5.A1 I58 2021

EDITOR
Christopher H. Roosevelt

AUTHORS
Ayşe Nur Akdal
Elton Barker
Anuja Dangol
Nicholas de Lange
Claudia Engel
Piet Gerrits
H. Murat Güvenç
Piraye Hacıgüzeller
M. Erdem Kabadayı
Vahakn Keshishian
Daniel Knitter
Turgay Koçak
Bernhard Ludwig
Dominik Lukas
Camilla Mazzucato
Osman Özkan
Jeroen Poblome
Michael Połczyński
Christopher H. Roosevelt
Amy Singer

COPY EDITORS
Lauren Davis
Jacob Chizzo

PUBLICATION COORDINATOR
Alican Kutlay

PUBLICATION ASSISTANT
Alara Demirel

BOOK DESIGN
Burak Şuşut, Fika

PRE-PRESS PRODUCTION
Fika

SYMPOSIUM COORDINATOR
Şeyda Çetin

CONTRIBUTING COORDINATOR
Ebru Esra Satıcı

PROJECT MANAGER
Buket Coşkuner

ANAMED DIRECTOR
Chris Roosevelt

PRINT
12. Matbaa
İbrahim Karaoğlanoğlu Cad. No. 35
34418 Kağıthane, Istanbul
T. +90 212 281 25 80
info@onikincimatbaa.com
Certificate No: 46618

PUBLISHER
KOÇ UNIVERSITY PRESS
Rumelifeneri Yolu, 34450
Sarıyer, Istanbul/Turkey
T. +90 212 338 10 00
kup.ku.edu.tr
kup@ku.edu.tr
Certificate No: 18318

FIRST EDITION
Istanbul, 2021

ISBN
978-605-7685-37-7

© 2021. This book has been published on the occasion of the 12th International ANAMED Annual Symposium "Spatial Webs: Mapping Anatolian Pasts for Research and the Public" that was held at the Koç University Research Center for Anatolian Civilizations (ANAMED), Istanbul, on 29–30 November 2017.

SPATIAL WEBS

Mapping Anatolian Pasts for Research and the Public

EDITED BY CHRISTOPHER H. ROOSEVELT

ANAMED
KOÇ UNIVERSITY
RESEARCH CENTER FOR
ANATOLIAN CIVILIZATIONS

KUP

Table of Contents

III. Mapping History

IV. Mapping Cultural Heritage and Identity

Preface

Koç University's Research Center for Anatolian Civilizations (ANAMED) in Beyoğlu, Istanbul was founded in 2005 to support research on the archaeology, art, heritage, and history of Anatolia and to help bring that research to the public through fellowships, exhibitions, publications, library resources, and symposia. After two years in the role of director of ANAMED, I took on the organization of the 12th International ANAMED Annual Symposium (IAAS)—traditionally a collaboration between ANAMED and Koç University faculty—both because of my interest in the subject and because I wanted to learn better through experience what running a symposium at ANAMED entailed.

The intent of the 12th IAAS in 2017 was to bring together scholars engaged in creating and using spatialized data resources for research and web-based data presentation and collaboration. The symposium aimed not only to showcase projects that incorporated GIS and other analytical tools in spatially-significant research into the past, but also to showcase projects that had engaged or would soon engage in spatializing collections of data for interactive web-mapping applications for researchers and the public alike. The event was popularly attended, and participants were cajoled into submitting revised and expanded versions of their presentations for review and publication, the result of which now lies before you.

ANAMED began publishing its own scholarly volumes in 2015 but began publishing IAAS volumes (in English and Turkish versions) only in 2018, with the proceedings of the 10th IAAS of 2015 (*Spolia Reincarnated: Afterlives of Objects, Materials, and Spaces in Anatolia from Antiquity to the Ottoman Era*). The 11th IAAS of 2016 (*Sacred Spaces and Urban Networks*) was subsequently published in 2019, and now the 12th IAAS of 2017 in 2020. Our aim moving forward is to publish the scholarly work stemming from each symposium within two years or so of its initial presentation. This goal is currently on track with the diligent work of the ANAMED Publications Office, now in the capable hands of Alican Kutlay, to whom I offer great thanks for his patient guidance, diligence, and schedule-keeping-nudges throughout the process.

For financial, administrative, and logistical support of the symposium and this resulting publication, many others deserve thanks, as well, beginning first and foremost with the Vehbi Koç Foundation (VKV) and Koç University (KU), the two organizations that make all things at ANAMED possible, along with Ömer M. Koç, VKV President Erdal Yıldırım, KU President Umran İnan, KU Vice President for Research and Development İrşadi Aksun, and College of Social Sciences and Humanities Dean Aylin Küntay. For logistical help with the symposium and its program, former ANAMED staff Şeyda Çetin and Özge Ertem were most instrumental, with the assistance of Ebru Esra

Satıcı, Ege Germen, and Müge Bulu Akar. ANAMED symposia and publications are not realized in bubbles but rely on the collaborative work of the entire ANAMED team; my sincere gratitude is extended to Buket Coşkuner, Duygu Tarkan, Mukaddes Gül, and Murat Türk, among others. Many thanks also to copyeditors, translators, designers, and draft readers, Lauren Davis, Jacob Chizzo, Duygu Çamurcuoğlu, N. Pınar Özgüner, Burak Şuşut, Gary Nobles, and Alara Demirel, as well as to KU Archaeology and History of Art undergraduate helpers A. Furkan İnan and Pelin Peker.

Publications and symposia are not realized without contributors, of course—their Most Valued Participants—and I remain especially grateful to all those who participated in the symposium—including moderators Ahmet Ersoy, Shirine Hamadeh, Hannah Lau, and Lucienne Thys-Şenocak—as well as to the contributors to this volume and those symposium participants who were unable to contribute to this volume. The latter includes Antonis Hadjikyriacou and Christos Chalkias, Nil Tuzcu, and Asım Evren Yantaç. Special thanks for his willing participation and both initial and *ex post facto* feedback goes to keynote speaker Elton Barker. Finally, for feedback on the initial, core idea of *Spatial Webs* and support throughout the process, a thousand and more thanks to Christina Luke.

Chris Roosevelt

Abbreviations

ACM	Association for Computing Machinery
ADS	Archaeological Data Service
AHRC	Arts and Humanities Research Council
ANAMED	Research Center for Anatolian Civilizations, Koç University
API	Application Program Interface
AVI	Audio/Video Interleaved
AWMC	Ancient World Mapping Center
BBC	British Broadcasting Corporation
BCE	Before Common Era
BMP	Bitmap Format
CAA	Computer Applications and Quantitative Methods in Archaeology
CALCS	Cross-Cultural After-Life of Classical Sites
CAMPOP	Cambridge Group for the History of Population and Social Structure
CC-BY	Creative Commons License
CERN	Conseil Européen pour la Recherche Nucléaire = European Council for Nuclear Research
CHGIS	China Historical GIS
CIDOC CRM	Conceptual Reference Model of the International Committee for Documentation
CLUSTERS	Consortium Linking Universities of Science and Technology for Education and Research
CRANE	Computational Research on the Ancient Near East
CSV	Comma-Separated Values Format
ÇRP	Çatalhöyük Research Project
DAAHL	Digital Archaeological Atlas of the Holy Land
DAI	Deutsches Archäologisches Institut = German Archaeological Institute
DARMC	Digital Atlas of Roman and Medieval Civilizations
DB	Database
DBF	dBASE Spreadsheet Format
DEM	Digital Elevation Model
DGPS	Differential Global Positioning System
DLR	Deutsches Zentrum für Luft- und Raumfahrt
DOC	Microsoft Word Processing Format
DWG	Computer-Aided Design Drawing Format
DXF	Autodesk Drawing eXchange Format

EMODnet	European Marine Observation and Data Network
ESRI	Environmental Systems Research Institute
EU-DEM	European Digital Elevation Model
FAIR	Findable, Accessible, Interoperable, and Reusable (principles)
FLV	Flash Live Video Format
FWO	Research Foundation Flanders
GBHGIS	Great Britain's Historical GIS
GDAL	Geospatial Data Abstraction Library
GDPR	General Data Protection Regulation
GeoJSON	Geographic JavaScript Object Notation
GeoTIFF	Geostationary Earth Orbit Tagged Image Format
GIF	Graphics Interchange Format
GIS	Geographic Information System
GLAM	Gallery, Library, Archive, and Museum (sector)
GPS	Global Positioning System
GRASS	Geographic Resources Analysis Support System
GUI	Graphical User Interface
HEROM	*Journal on Hellenistic and Roman Material Culture*
HGIS	Historical Geographic Information System
HTR	Handwritten Text Recognition
IAAS	International ANAMED Annual Symposium
IDE	Integrated Development Environment (Python)
IFEA	Institut Français d'Études Anatoliennes = French Institute of Anatolian Studies
IIIF	International Image Interoperability Framework
INSPIRE	Infrastructure for Spatial Information in Europe
ISC	Istanbul Studies Center, Kadir Has University
KMKD	Kültürel Mirası Koruma Derneği = Association for the Protection of Cultural Heritage
KML	Keyhole Markup Language and File Format
KOPAL	Konya Basin Palaeoenvironments Project
KU	Koç University
LAÇ	Living Archive of Çatalhöyük
LOD	Linked Open Data
LODE	Linking Open Descriptions of Events
LPWG	Linked Pasts Working Group

MedArchNet	Mediterranean Archaeological Network
MEDMAPS	Mapping Mediterranean Lands
MHATNF	Maritime History Archives of the Turkish Naval Forces
MIT	Massachusetts Institute of Technology
MOV	QuickTime Movie Format
OCHRE	Online Cultural and Historical Research Environment
OCR	Optical Character Recognition
ODATE	Open Digital Archaeology Textbook Environment
ODF	Open Document Format
OpenAIRE	Open Access Infrastructure for Research in Europe
PDF	Portable Document Format
PostgreSQL	Postgre Structured Query Language
PST	Primary, Secondary, Tertiary System
QGIS	Quantum Geographic Information System
RDF	Resource Description Framework
RMSE	Root Mean Square Error
RTF	Rich Text Format
SIGCHI	Special Interest Group on Computer-Human Interaction
SIIS	Sagalassos Integrated Information System
SLLP	Stanford Literary Lab Pamphlet
SRTM	Shuttle Radar Topography Mission
TAYGIS	Türkiye Arkeolojik Yerleşmeleri GIS = Archaeological Settlements of Turkey GIS
tDAR	The Digital Archaeological Record
TEI	Text Encoding Initiative
TIF	Tagged Image Format
TPI	Topographic Position Index
UML	Unified Modeling Language
UNESCO	United Nations Educational, Scientific, and Cultural Organization
URI	Uniform Resource Identifier
VKV	Vehbi Koç Foundation
XLS	Microsoft Excel Spreadsheet Format
XLSTAT	Statistical Software for Excel
XML	Extensible Markup Language

PAPERS

I

Mapping and Annotating the Past

Location Matters: Disciplinary Diversity and Community Collaboration in Digitally Mapping the Past

Christopher H. Roosevelt

Introduction

Recent decades have witnessed an explosion in digital and spatial technologies that allow facile management, interactive display, and analysis of diverse types of data. Datasets informing past demographic, economic, political, military, religious, and other phenomena are just a few examples, and their querying and analysis enables exploration of settlement and communication networks, material distributions, social and political boundaries, and census patterns, for example.

This phenomenon is part and parcel of developments in digital humanities, broadly defined, that have leveraged text, image, and geographic resources in new ways, taking advantage of concepts deriving from data-science approaches and advances in digital data storage, manipulation and analysis, and presentation. At the same time, it is a consequence of the increasing user-friendliness and mainstream usage of maps and mapping, and Geographic Information Systems (GIS), in particular, with its inherent advantages for spatially significant investigations. It was only a matter of time before the convergence of related trends resulted in the emergence of "historical GIS" (or HGIS) and, slightly more recently, "spatial history," "spatial humanities," and even "geohumanities;" these new approaches, or at least new terms, appear ever more frequently in article, book, and series titles, if not also project and grant proposals (e.g., Gregory and Ell 2007; Bodenhamer, Corrigan, and Harris 2010; White 2010; Dear et al. 2011; Gregory and Geddes 2014). In all such approaches to understanding the past via digital tools for spatial data management and analysis, the idea that *location matters* is inherent in understanding places across space and through time, their connections and individualities.

It is based in part on this idea—*location matters*—that Koç University's Research Center for Anatolian Civilizations (ANAMED) convened its 12th International ANAMED

Annual Symposium (IAAS) on the subject of "Spatial Webs: Mapping Anatolian Pasts for Research and the Public" in December 2017. The primary aim of the symposium was to provide a forum to share how particular digital mapping and spatial analysis initiatives were being or could be employed for research and the public, demonstrating methods used in data collection and their spatialization, as well as the diversity of ways in which underlying datasets are being or could be managed, visualized, and interacted with, especially online or via web-based platforms. That is, the primary aim was to explore how "spatial webs" could be leveraged for research and public presentation, and in particular with past Anatolian civilizations in mind.

The idea for the ANAMED symposium and this subsequent volume did not arise *ex nihilo*, of course, and could even be considered very late to the party, following in the footsteps of the general developments mentioned above as well as then relatively recent initiatives, conferences, and publications, as it did. Just concluded in 2016 was the University of Lancaster's European Research Council project "Spatial Humanities: Texts, GIS, & Places" (Spatial Humanities 2016), led by Ian Gregory, a pioneer in HGIS (Gregory and Ell 2007; Gregory and Geddes 2014). Then-recent conferences included "Mapping the Past: GIS Approaches to Ancient History," the first conference organized by the Ancient World Mapping Center at UNC Chapel Hill (AWMC 2016), established earlier by Richard J. A. Talbert, well-known historian of ancient mapping and mapper of ancient history (Talbert 2000). Stanford University's Center for Spatial and Textual Analysis and Spatial History Project, also, had produced much work, with online publications presenting results since 2008 (e.g., White 2010). The keynote speaker at the 2016 AWMC conference and founding director of that center, Tom Elliot, helped develop and still directs the Pleiades project (Pleiades, n.d.). Pleiades began as an updated online version of Talbert's (2000) *Barrington Atlas* and has expanded significantly to become (one of) the largest open, collaborative, and online gazetteer(s) of ancient places in the Mediterranean world yet developed.

To be sure, many other projects had already not only engaged in HGIS or spatial history but also leveraged the web—as an interactive platform for presenting, sharing, and working with data—to enable research collaborations and to bring mapping projects to the public. Many of these, too, focused on the Mediterranean world, perhaps owing to the richness of its historical and archaeological record. Pioneering in this realm were Roman historian Walter Scheidel's ORBIS, an online, multimodal, transportation model of the ancient Roman Empire (Scheidel and Meeks, n.d.), followed by other Roman route-plotting portals based on ancient itineraries (e.g., Voorburg 2011; DARMC, n.d.; Weber, n.d.). Other online and spatial projects have missions that range from providing scanned versions of archival maps (e.g., Earthworks, n.d.; MEDMAPS 2005) to presenting atlases of specific periods or regions (some of which share data from Pleiades; e.g., Voorburg 2014; Digital Crete, n.d.) and to enabling the creation of custom maps showing ancient places (e.g., Antiquity À-la-carte, n.d.). Of such projects, the Digital Archaeological Atlas of the Holy Land, established by Thomas E. Levy and Stephen H. Savage (DAAHL, n.d.), might be one of the broadest reaching. In addition to scanned archival maps, atlases, and custom maps, it includes also the Mediterranean Archaeological Network, designed to be a "portal-science" platform for collaboration and sharing of research conducted across the Mediterranean (MedArchNet, n.d.). As a platform for collaboration, it could be compared with the data-aggregator and Pleiades-linked platform of Pelagios Commons and its data exploration tool Peripleo (Pelagios, n.d.; Peripleo, n.d.; Barker, this volume).

Online geospatial projects focusing (primarily) on (parts of) Anatolia had been available for some time before the 2017 ANAMED symposium, as well. Research-aggregating and collaboration-inspiring platforms included David Schloen and Sandra Schloen's OCHRE project at the University of Chicago (OCHRE, n.d.; Schloen and Schloen 2012). By that time, German Archaeological Institute (DAI)-led research in Anatolia, too, was already being catalogued into the online archives and linked gazetteers of iDAI.world, a robust repository and geospatial database platform for sharing and exploring DAI-produced data (iDAI, n.d.). Still other projects had begun to aim for similar goals, even if online access had yet to be achieved (e.g., the CRANE project of the University of Toronto; Harrison 2018). Some projects concerned all Anatolia, such as the Archaeological Settlements of Turkey Project, which catalogues archaeological settlements in Anatolia as well as their harm or destruction (TAYGIS, n.d.), and the initiative of Pascal Lebouteiller at the French Institute of Anatolian Studies to scan and make available its rich archive of historical maps (IFEA Map, n.d.). Other projects focused only on Istanbul, such as Nil Tuzcu's presentation of a variety of archival and recent map data in her interactive Istanbul Urban Database (Tuzcu, n.d.) or Holger Klein's Istanbul Documentation Project (Klein, n.d.). This is not to mention, of course, the several projects described in this volume that had previously developed similar mapping and/or online initiatives for various Anatolian geographies and/or subjects (e.g., de Lange; Keshishian; Mazzucato, Engel, and Lukas; and Połczyński and Singer).

This brief and selective overview of conference, project, and publication precedents to the 2017 symposium at ANAMED leaves out many seminal projects (e.g., Harvard's China Historical GIS project [CHGIS, n.d.] and national HGIS platforms like that of Great Britain [GBHGIS, n.d.]); it is meant only to demonstrate the rich history of relevant, previous work, including that focusing on Anatolia. Nonetheless, even if launched from the shoulders of such predecessors, the "Spatial Webs" symposium of 2017 appears to be the first dedicated to mapping projects and online mapping portals as they applied to Anatolian landscapes and history, and thus it offered an opportunity to compare notes among researchers engaged in related research and dissemination trajectories.

The outcome is the diverse set of chapters presented here, arranged into four sections. Following this introductory chapter and one highlighting the increasing significance and potential of annotation, with broad-ranging relevance and resonances across all subsequent chapters, the three remaining sections include examples of mapping in related approaches to Anatolian pasts, somewhat artificially grouped into sections on Archaeology, History, and Cultural Heritage and Identity (Fig. 1). Some chapters present research- and web-based collaboration and public sharing platforms, others present research projects preparing for or ready to serve as self-standing spatial-data exploration portals or to be linked with broader integrative platforms. Rather than review each chapter in customary introductory fashion, the remainder of this chapter will instead discuss contributions to the volume along the lines of a few general trends, attempting to tug on some interesting threads, to muse on a few themes.

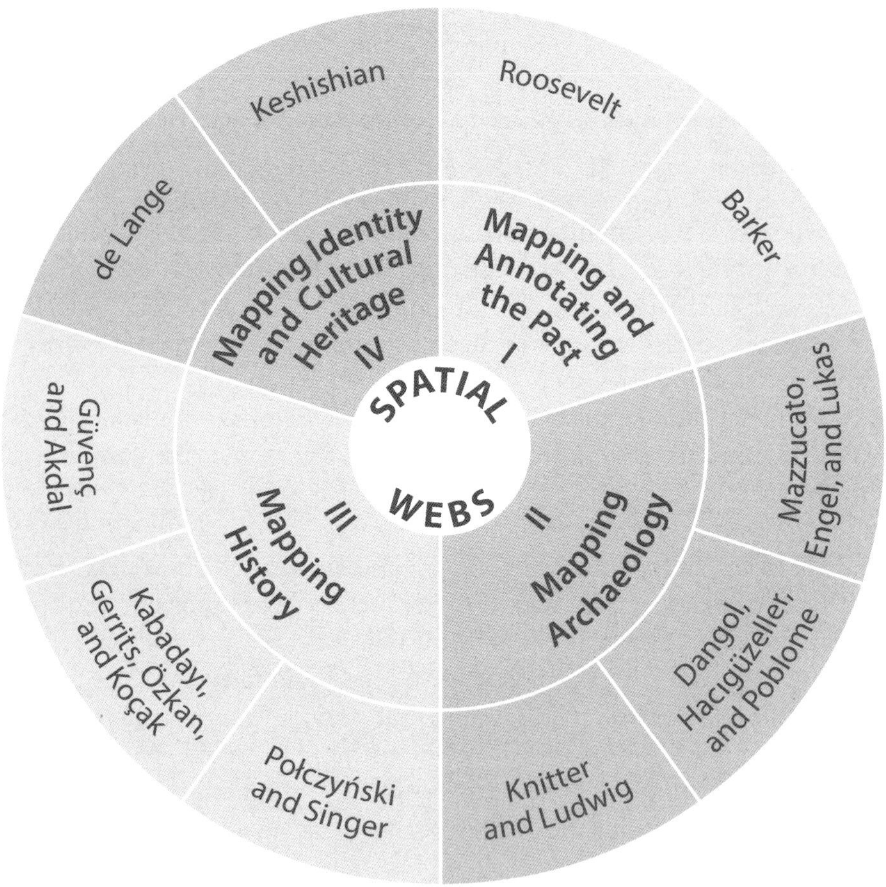

Fig. 1. Chapter authors and volume sections in this *Spatial Webs* volume.

Disciplinary Diversity

In reviewing their approach to creating an online platform for sharing research on the Ottoman world, Połczyński and Singer (this volume) suggest that "fields of research within the humanities are not equally digital and digitized," a statement echoed by other contributors to this volume (e.g., Kabadayı et al.) and similarly applicable to the social sciences. Just as valid would be the observation that such fields are not equally spatial and spatialized. In this volume these differences of *disciplinary location* are perhaps best represented by differences among archaeologists and historians engaged in digital research, here defined simply as research taking advantage of computational power (White 2010, 1; Blevins 2014, 23).

The use of computers in archaeology began early enough to warrant critical review articles beginning already some fifty years ago (e.g., Cowgill 1967; Whallon 1972; see also Wilcock 1989; Richards 1998). The Computer Applications and Quantitative Methods in Archaeology (CAA) group has convened since 1973 with annual proceedings (e.g., Wilcock 1973), and textbooks and edited volumes providing practical guidance began to appear only slightly later (e.g., Doran and Hodson 1975; Richards and Ryan 1985; Lock and Wilcock 1987; Reilly and Rahtz 1992; McPherron and Dibble 2002; Lock 2003). GIS has a long history of use in archaeology, as well, with significant advances dating to the 1980s (e.g., Allen, Green, and Zubrow 1990; Gaffney and Stančič 1991; or Lock and Harris 1991). Compare this with the situation in history: early calls urged historians to harness computational power around fifty years ago (e.g., Silveira 2014, 41, citing Ladurie 1973, 11–14); the use of GIS in historical research began in the 1990s (Gregory, Kemp, and Mostern 2001; Knowles 2002), matured in the 2000s with the emergence of HGIS and then spatial history (Gregory and Healey 2007, 642–44; White 2010), with spatial humanities (a correlate to digital humanities) appearing in the 2010s (Gregory and Gregory and Geddes 2014).

Computers and digital mapping approaches appear to have been introduced to archaeology and history around the same time, then, yet one could argue that they were embraced by archaeologists both more quickly and more fully. This might be linked to the nature of the discipline, whose reliance on spatial recording dates back more than one hundred years. Gillings, Hacıgüzeller, and Lock (2019, 1) emphasize that "[f]rom the very beginning of archaeological practice, maps (and plans) have been one of the discipline's most fundamental tools." A longer tradition of attention to spatial significance might also be related to the embodied nature of archaeological fieldwork, requiring engagement with the places and spaces of past human activities more physically and directly than traditional historical research, as well as to connections between archaeology and earth sciences.

The more embedded tradition of digital research in archaeology might be reflected in the relative frequencies with which archaeologists versus historians in this volume refer to things "digital," in general, versus the relatively new trends of

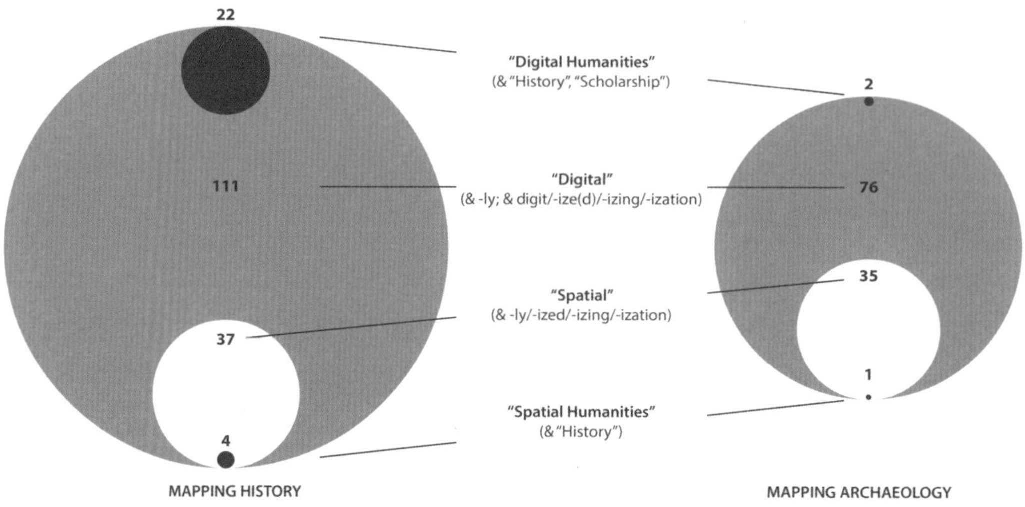

Fig. 2. Frequency comparison of "digital" and "digital humanities" (including also "digital history" and "digital scholarship") and "spatial" and "spatial humanities" (including also "spatial history") in the chapters that make up the "Mapping History" versus "Mapping Archaeology" sections of this volume. Graduated circles are illustrated proportionally.

"digital humanities" and "digital history" (Fig. 2). Among chapters in the history versus archaeology sections of this volume, for instance, historians use "digital" and its variants 111 times, 46% more than do archaeologists (76 times), while historians also highlight "digital humanities," "digital history," and/or "digital scholarship" 22 times, 1000% more than do archaeologists (only 2 times). Do archaeologists feel less need to use these relatively recent terms because they have been more comfortably "digital" for a longer time? Even "digital archaeology" and its variants—recent terms used by archaeologists usually to refer to field-data collection workflows as much as data storage, management, analysis, and display, which have long been digital—appear only 7 times among the archaeology chapters.

Even as *disciplinary location* affects the degrees and ways in which digital mapping approaches have been or are embraced, certain commonalities are apparent, as well. Of most relevance here, of course, are conceptions of *location* itself, commonalities perhaps reflected in the more balanced pattern of mentions of things "spatial," in general, versus "spatial humanities" and "spatial history" among disciplines represented in this volume (Fig. 2). Geographers (and particularly historical geographers) and many archaeologists among the readers of this volume will note that spatial approaches can hardly be considered new in disciplines in which interpreting evidential distributions of past and present societies has been a core activity since the early twentieth century. As James Mellaart (1970, 55), a prominent Anatolian archaeologist of that century, put it, "... unless one knows the distribution of a phenomenon, one knows nothing worthwhile about it." What is relatively new, however—even to these fields—is the increasing ability to collate, display, and analyze such data efficiently and effectively through the use of digital tools, foremost among them GIS, and, more recently, the ability to disseminate them through the web for presentation and collaboration.

Also relatively new, perhaps, is the trend referenced generally as the "spatial turn" (e.g., Döring and Thielmann 2008; Warf and Arias 2009; Mitchell 2011; Barker, this volume), now pervasive in archaeology, history, and cultural heritage literature. The trend towards increasing attention to spatial questions and approaches is certainly influenced by the development of spatial technologies, but it draws much earlier and more significantly on Braudelian framings of landscapes and time, as well as from Lefebvre, Tuan, and other recent scholars' conceptualizations of place and space as both human constructs, the productions of particular people at particular locations and times, and influential agents in and of themselves (Braudel 1949; Lefebvre 1974; Tuan 1977; Ethington 2007; White 2010; Barker, this volume). While archaeology may have embraced digital mapping approaches and tools (e.g., GIS) more quickly and fully than did history, these fields are currently in similar places concerning critiques of previously colonial, detached, positivist, and other narrow, pseudoscientific, or biased approaches to spatial interpretation (Cosgrove 1984; Tilley 1994; Thomas 1993, 2004; Paliou, Lieberwirth, and Polla 2014; Gillings, Hacıgüzeller, and Lock 2019). Concerning a more critical spatial history, for example, Silveira (2014, 33), among others, argues for a "conceptual framework to deal with the notion of space in history in order to better understand the relations between space, individuals and societies across time." Some have attempted just this, highlighting links between the spatial turn and globalization (Middel and Nauman 2010); others have tried to develop new definitions of history and time themselves, as inherently spatial (or platial) concepts (e.g., Ethington 2007). No consensus framework has been reached, however, and most scholars go on approaching spatial research questions with the theory and methods—as well as the technological tools—they think best fit the needs of particular studies.

Thus, while in this volume historians Kabadayı and colleagues may use terms like HGIS, digital/spatial history, or digital/spatial humanities to mean georeferencing of historical maps to modern Cartesian systems for digitization and analyses, Połczyński and Singer, Güvenç and Akdal, de Lange, and Keshishian may do so to include the spatialization of literary or archival sources for online dissemination with dynamic cartographic display and querying. Similarly, Mazzucato, Engel, and Lukas, as well as Dangol, Hacıgüzeller, and Poblome, may be most concerned with the standardization, spatialization, and online presentation of data in online, "living" databases of decades of digitally and geospatially sophisticated archaeological work, whereas Knitter and Ludwig apply relatively new spatial analytical methods to recent archaeological data using open-source methods as an example of reproducible research. Despite such diversity, all authors—whether from the field of archaeology, history, or cultural heritage—are concerned with understanding records of human activity and mobility across diverse past landscapes, engaging with increasingly complex data science and spatial methods to preserve and present component and resultant data, often via the capabilities of the web.

Researcher Shape and Collaboration

Should up-to-date, cutting-edge scholars then aim for full mastery of data-science and/or GIS methods in addition to that of the particular domain science(s) from which they approach spatial questions about past societies (e.g., archaeology, history, cultural heritage)? That is, does the application of data-science and GIS require deep and special methodological training in coding and GIS? Or do attempts to leverage potentially big data from the digitization of archival records and maps require specialization in Optical Character Recognition (OCR), Handwritten Text Recognition (HTR), or related technologies? In essence, have digital, online, and spatial tools advanced in ways so differently from past research tools that they necessitate separate, specialist, and deep training in addition to theoretical and subject-specific topics? A quick possible response: perhaps no more than other types of newly developed research tools have required specialist training(s) in the past. Avoiding specific debates here such as that between GI*System* (a tool) versus GI*Science* (a discipline) (e.g., Goodchild 1992), it is good to consider such questions in the context of recent comments about the training of future scholars.

Knitter and Ludwig (this volume) argue that, in order to contribute most meaningfully and openly to our disciplines, the times demand serious digital approaches to data, its storage, analysis, and open dissemination, as well as the transparency of full analytical processes and interpretations. Barker (this volume), too, emphasizes FAIR principles and argues that open approaches to data, metadata, and analyses in effect help preserve democratic traditions in ensuring the transparency of the intellectual processes through which connections are made and conclusions are drawn. Few would argue with such statements. One primary question that arises, of course, is by what means such ends are fulfilled: by individual scholars, by groups of scholars, or combinations thereof.

Discussion revolves around the concept of "researcher shape," wherein T-shaped researchers have a wide breadth of coverage (the horizontal bar of the T) as well as a deep knowledge (the vertical bar) in the traditional theoretical and subject-based approaches to a particular field (e.g., Late Ottoman history). Knitter and Ludwig (this volume, Fig. 2) argue that the "*modern* researcher... needs expanded expertise not only in her discipline, but also in a second field, namely that of computing skills and reproducibility" (emphasis original). The deep knowledge of the second field—which could be any of the specialisms mentioned above—adds a second vertical bar to the

T, thus producing a Π, with the "Π-shaped researcher" concept first attributed to data scientist Alex Szalay (Faris et al. 2011, 215). Thus a Π-shaped Late Ottoman historian might have deep expertise in both that primary field and GIS, HTR, or some other digital or data-science approach. Few exceptions are taken with encouragement of diversifying one's approaches and achieving the gratification of standing with both legs planted firmly: one in a primary and the other in a secondary field. Yet understandable concern arises about the availability of time for such dual training (at the expense of primary field training) and the possibility that a secondary field becomes privileged over a primary.

In discussing the use of GIS in history, for example, Silveira (2014, 43) suggests that "the emphasis should be put not on HGIS, an area characterized by the application of a technology, but on spatial history, a field where space receives a special attention," especially in addressing the *Annales*-type questions and problems that have been at the forefront of landscape-grounded research since Braudel. In comments that apply to archaeologists and cultural heritage experts as much as they do historians, Gregory (2014, 1) avers that, "[t]he challenge for digital historians ... is to do good history that combines the computer's ability to search and summarize, with the researcher's ability to interpret and argue." He goes on to suggest that all historians need to do in navigating current digital and spatial trends, their challenges and opportunities, is "to remember that they are first and foremost historians and that historians fundamentally are in the business of taking complex, incomplete sources that are full of biases and errors, and interpreting them critically to develop an argument that answers a research question" (Gregory 2014, 1). In short, the first leg of the Π should be kept stronger and longer, as it defines the primary identity of the researcher as well as research questions and holistic approaches.

From the perspective of primary fields in the social sciences and humanities, it is instructive to look to data science itself, where internal critiques suggest other, perhaps more achievable goals that can lead to more sustainable futures. In presenting the results of their ethnographic study of a university data-science center that serves as an incubation hub for collaborations with domain sciences, Fiore-Gartland and Tanweer (n.d.) suggest that a network of collaborators with diverse and overlapping areas and levels of expertise is a more promising concept. That is, rather than producing bevvies of Π-shaped researchers, the aim should be to bring together interested groups of T-, Γ-, M-, and Π-shaped researchers, each of whom brings a particular combination of expertise to the table (Fig. 3). In summary, while those with the interest, capability, and time could become independent Π-shaped researchers, many scholars could achieve similar if not greater research outcomes by developing above-average knowledge depth in a secondary data or spatial science, yet also collaborate with a research group, each member steeped in complementary areas of skill and expertise.

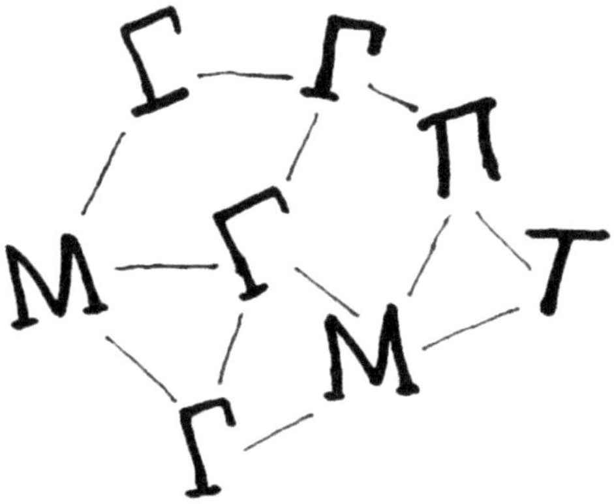

Fig. 3. What community-level data science (or your average archaeological project) might look like, including a collaboration of T-, Γ-, M-, and Π-shaped researchers (image by Brittany Fiore-Gartland under ⓒ BY-NC license).

This conclusion comes with little surprise to anyone who has tried to integrate data science or spatial analyses with social sciences and humanities. As the authorship of most of the chapters in this volume demonstrates, large digital and spatial research projects are almost always collaborative, many because they must be (Połczyński and Singer, this volume). While a "traditional historian" (however that might be defined) might still work primarily alone, a digital or spatial historian rarely does so (White 2010, 1; Silveira 2014, 30–31, 42). Good archaeological work and the data-science and spatial analysis of its results are almost never solitary. The network of differently shaped collaborators advocated by Fiore-Gartland and Tanweer (n.d.), in fact, is a fairly accurate representation of a traditional archaeological (and now also digital or spatial history and cultural heritage) team. As in all team-based ventures, good communication is an essential component of successful collaborative efforts, especially when increasingly complicated technologies and deep knowledge within particular domains offer increasing potential for lack of understanding among team members (Faris et al. 2011). Communication is even more important when the definition of research collaboration expands beyond single projects, towards inter-project data collectivizing or aggregating, and inter-project data analyses. Here, just as important as interpersonal communication is the accurate communication and connectivity of datasets, a topic of growing significance for the fields represented in this volume.

Data Harmonization, Linking, and Openness: Towards Community-Level Science

The number of social science and humanities researchers willing to collaborate and share not only research outcomes but also underlying data and analytical workflows is increasing, a trend directly enabled by the web (version 2.0) as a platform of collaboration (Kansa, Kansa, and Watrall 2011; Shanks and Whitmore 2012; Galeazzi and Richards-Rissetto 2018; Wright and Richards 2018). Easy, internet-enabled interpersonal and data communication has allowed like-interested researchers to form communities of scholars working on singular projects or sets of interlinked projects more fluidly and effectively than possible otherwise. This has led to the open-ended, dynamic, and powerful concept of a "scholarly commons" not based on printed material (e.g., White 2010, 1; Barker, this volume). Well-known examples of such collaborative initiatives include Pleiades (Pleiades, n.d.), the "community-built gazetteer and graph of ancient places" mentioned above and in many chapters of this volume, as well as Pelagios Commons and its component parts (Pelagios, n.d.; Barker, this volume). Such projects explicitly emphasize their "community" nature, relying on "expert sourcing" of data (if not also workflows) from a community of decentralized expertise. To preserve the intercommunicability of such decentralized projects, extra care must be taken in data hygiene protocols that ensure internal as well as external harmonization.

Internal harmonization of data and metadata refers here to the production and curation of new data just as it does to legacy data (e.g., Gregory and Healey 2007, 643–44), and it involves their standardization or normalization according to structured data storage protocols. With respect to new digital data deriving from archival texts and place name gazetteers, the historical and heritage projects described by Połczyński and Singer, Güvenç and Akdal, de Lange, and Keshishian, for instance, involve the careful structuring of previously unstructured and newly extracted data. Połczyński and Singer, de Lange, and Keshishian then essentially georeference that textual data, enabling its geographic display. The structuring of demographic data from archival documents allows Güvenç and Akdal to achieve similar spatialization goals following statistical analysis. Kabadayı and colleagues, alternatively, work directly from archival maps to produce structured data for GIS analyses. Borrowing from the concept of "data mining" now commonly associated with big datasets, they engage in "map mining" to digitize entities from georeferenced archival map sources, enabling point-to-point route and other analyses.

The internal harmonization of archaeological data described in this volume involves long-term or legacy data. Mazzucato, Engel, and Lukas trace the twenty-five-year history of the Çatalhöyük Research Project's collection of diverse data in different formats, their harmonization, and, more recently, the web-sharing of a "Living Archive" for research and public access. Dangol, Hacıgüzeller, and Poblome highlight challenges in transforming archival analog and digital data for use in a new spatially capable system: the Sagalassos Integrated Information System. Both of these archaeological projects engage also in ensuring uniformity of terminologies used in structured datasets, following International Organization for Standardization (ISO) guidance, and mapping them to the Conceptual Reference Model of the International Committee for Documentation (CIDOC CRM), or variants thereof, in order to enable semantic-web interaction.

External harmonization here refers to the preparation of data for just this sort of linking and semantic intercommunicability. To allow web-connected computers to "read" them, data need to be recorded and stored in ways that are meaningful to computers; that is, they need to be machine readable. Both Barker and Mazzucato, Engel, and Lukas review what such data legibility involves, including the tagging of data entities with Universal Resource Identifiers (URIs), essential for preventing ambiguity, and the expression of relationships between them according to Resource Description Framework (RDF) triples. Entire databases can be "semantically mapped" in this way, but so too can selected bits and pieces of originally non-structured data, such as particular words in texts and places on maps. This is, in essence, the process Barker describes in detail concerning annotation. By annotation, Barker means a type of tagging whereby specific entities—usually people, events, or places—occurring in a specific digitized medium (e.g., a text or image) are annotated with whatever comments or metadata is desired and enhanced with links to external datasets (e.g., texts, images, or maps). As long as such external datasets are *openly accessible*, semantic linking—now relatively easily achieved thanks to Pelagios Commons' powerful annotation tool Recogito (Recogito, n.d.)—allows computers to find all other instances of annotated entities, thus immediately revealing connections and patterns that may otherwise remain obscure.

The open accessibility of datasets can refer to the online archiving of entire research packages on repository websites like Zenodo or GitHub or sites specifically catered to hosting archaeological data, like Open Context (n.d.), the Digital Archaeological Record (tDAR, n.d.), and the Archaeological Data Service (ADS, n.d.); they can contain raw and refined datasets, analytical workflows, outcomes, and even full narrative write-ups. Knitter and Ludwig (this volume) emphasize how such open practices of archiving ensure transparency and reproducibility of results, principles foregrounded also by Barker. Such data archiving sites also offer the possibility of preserving the "intermediate data" described by Połczyński and Singer, the in-process research comments and notes that demonstrate the decision-making pathways that are often left unpreserved following publication.

The open accessibility of datasets can also refer to the type of external harmonization that enables the linking of data described above as well as to some level of Creative Commons licensing that allows for connectivity and reuse. With linking and licensing, datasets can participate in Linked Open Data (LOD) networks. With respect to this volume, de Lange's "Mapping the Jewish Communities of the Byzantine Empire" project links relevant place-names directly to Pleiades' place resources via RDF triples, and one could imagine the "Turkey Cultural Heritage Map" described by Keshishian being similarly linked. These linkages primarily concern place data, and yet LOD can involve far more than the annotation and linking of just places, including whole bibliographies and groups of databases as well as online repositories of, for example, zooarchaeological materials (Kansa 2014) or coins (Heath 2018; for further examples, see Elliott, Heath, and Muccigrosso 2014).

Returning now to the importance of communities of collaboration referred to above (Fig. 3) and the open accessibility of increasing numbers of projects and datasets, one could posit that their *decentralized location* is essential for success. The web-enabled decentralization of many digital projects allows for their growth and sustainability: they are not necessarily tied to one institution, department, group, individual, or location, and can thus draw from far broader pools of expertise and resources to achieve what is referred to as "community-level science" (Fiore-Gartland and Tanweer, n.d.; see also Snow and Dibner 2016, 73; Slater, Huxster, and Bresticker 2019, 250). Acknowledging the challenges that arise from the same circumstances (especially sustainability, see below), this configuration of large digital projects as community consortia appears to be ever more common, as suggested by the increasing number of projects on sites like Zooniverse.org or by Google's entry into the arena with its Dataset Search tool. With respect to the history or culture-oriented datasets described in this volume, community-level collaboration is underlined by the formation of the Linked Pasts Network (Grossner et al. 2017) associated with Pelagios, and it is what Pleiades and Pelagios are already engaged in for the ancient Mediterranean world, what OpenOttoman is pioneering for the Ottoman world, and perhaps also what AriadnePlus and the World-Historical Gazetteer, for example, hope to bring together for European archaeology and world history (AriadnePlus, n.d.; WHGazetteer, n.d.).

Looking Forward: Towards a Conclusion

As recounted above, one aim of the 2017 symposium was to provide a forum for scholars to share current approaches, results, and general experiences in mapping Anatolian pasts for research and public dissemination. The theme of community-level collaboration highlighted in this chapter was already emphasized in the symposium itself, wherein the importance of bringing together and linking people to form research communities was understood to be equal to that of linking datasets. A clear success could be marked soon after the event, when symposium participants began joining the Pelagios Commons community, for instance, to explore the utility of Recogito for mapping and linking entities extracted from Ottoman archives (Hadjikyriacou 2018a, 2018b, 2019). Otherwise, the symposium succeeded in doing what many such events often do (and what Barker suggests many graphs [and maps, too] should do by nature): provoke discussion and stimulate new ideas. Among the many topics touched upon in the symposium that remain less resolved or explored than otherwise are the use of alternative forms of mapping and visualization of spatially significant data, the sustainability of community-level collaborative ventures, and specific ideas for "spatial webs" of Anatolian pasts. Looking forward, and admittedly only skimming the surface, each of these subjects offers significant opportunities and challenges.

Most contributions to this volume invoke traditional cartographic conceptions of maps and ways of mapping, leaving relatively unexplored alternative forms of the same that move well beyond two-dimensional displays of spatially anchored data. Phenomenologically influenced forms of mapping that record personal, embodied, and perhaps even psychological experiences now take advantage of the social and participatory capabilities of Web 2.0 and mobile applications, for instance; examples of experimental, counter-cartographic, and other mapping approaches abound in recent edited volumes focusing on geography (Thompson 2008), the social sciences and humanities, in general (Dear et al. 2011), and archaeology, in particular (Gillings, Hacıgüzeller, and Lock 2019). For heritage and other historical mapping efforts, the concept of deep mapping (and deep geography) might be especially rich (Bodenhamer, Corrigan, and Harris 2015), with its potential for embedding previously unattainable levels of richness and qualitative depth into geographically framed, observational windows (i.e., maps), enabling contextualized yet non-quantifiable connections across space and time.

The chapters of this volume provide additional examples. Barker's Hestia project, in part a network analysis of an ancient text, for instance, highlights ways of envisioning geographic interactions from the perspective of a particular author in a particular time and place. Similar "network maps" can be considered concept maps that highlight geographic connection and significance more than Cartesian coordinates or Euclidean space. The 2018 update to the "Turkey Cultural Heritage Map" described by Keshishian pushes it in the direction of deep maps by embedding multimedia testimonies directly into its online map. The Hrant Dink Foundation's recent and related KarDes mobile app focuses on just this, with its "Multicultural Memory Tours" of Istanbul (Hrant Dink Foundation, n.d.). It can be expected that such alternative approaches to mapping Anatolia's deep and diverse pasts will become more common in the future.

A second unresolved theme related to digital and spatial research into the past is project sustainability, and, in particular, that relating to data institutionalization and the academic processes associated with it. By data institutionalization, I mean the long-term sustainability of data accessibility, as well as the management of content versioning and file-format transcoding. De Lange makes clear that the sunsetting of funding for "Mapping the Jewish Communities of the Byzantine Empire" meant that its data was no longer to be updated, yet Keshishian describes a user-sourced method of updating through online comments. Will informal and online curatorial communities be sufficient? De Lange, as well as Knitter and Ludwig, archive project data and their entire project, respectively, on Zenodo, and one hopes that it and similar repositories (e.g., GitHub, Zooniverse, etc.) will only become increasingly robust over time. Połczyński and Singer's and others' pleas for long-term, durable data platforms, however, point squarely in the direction of libraries, and these in particular, together with other institutional repositories within the gallery, library, archive, and museum (GLAM) sector, more generally, present encouraging prospects. University libraries—with their already robust experience in database management and connection to scholarly communities—have the necessary strong foundations for sustainable curation and presentation of data, and many are increasingly interested in providing institutional repositories for faculty research, as well. And even already well-established archaeological data repositories like Open Context, tDAR, and the ADS are associated with university library systems.

How a university (and academia in general) recognizes researcher effort in developing content for and platforms of data-heavy digital projects is another issue directly affecting sustainability, particularly that of projects committed to participating in openly accessible, online, community-level initiatives. Scholarly communities cannot survive without clear mechanisms for rewarding research and service to the discipline (e.g., promotion), and digitally collaborative communities are no different. Połczyński and Singer rightly recognize that existing mechanisms need to be revised if open-data digital projects are to be fruitful and multiply. Keshishian's simultaneous publishing of printed volumes in association with online maps offers one possible avenue, although it raises a different set of questions concerning sustainability. Other possible solutions include phased or partial openness of data, evaluating publication of data online similarly to traditional publication evaluation (e.g., Kansa and Kansa 2014), and, following the recommendations of the Center for Open Science (n.d.), rewriting the way academic credit is granted for publishing both data and analyses, if not also the entire peer-review process. Time will tell how such proposed revisions to core academic traditions play out, yet the more data is published in openly accessible ways, the more likely it will be that some combination of these solutions gains traction.

The final topic that remains largely unexplored and thus unresolved by the 2017 symposium and this volume relates to one last desired outcome: discussion of specific principles and proposals for how multi-period and multi-scalar datasets of diverse types but common geographies—particularly those defined by Anatolian civilizations—could be integrated in ways that leverage their diachronic potential for longitudinal research across time and space. ANAMED itself is defined by a mission aiming to foster the disciplinary and chronological cross-pollination of research united by geography, recognizing the concept of landscape as palimpsest and that location, natural constraints, and the accumulation of past cultural residues affect the ways in which societies inhabit and respond to their physical and cultural environments.

The assembly of papers in the symposium and the chapters of this volume represent steps towards this goal, and the plethora of good case studies they review show that spatially linked, online data and mapping portals are an excellent mechanism to enable that for which ANAMED strives, looking both forward and backward through the lens of geographically-rooted explorations. With the importance of collaboration, data harmonization, and open linking clearly integral to the establishment of communities of like-interested scholars, all that is missing at present is the dedicated project or platform through which an Anatolian cluster of mapping data and research could contribute to broader LOD communities. Should ANAMED expand to fill this void, bringing broad swaths of spatially significant data and research to a "Linked Anatolian Pasts" platform for community-level research as well as public dissemination and education? It is a hefty challenge, but a clear goal for its future.

Works Cited

ADS. n.d. Archaeology Data Service. Accessed January 13, 2020. https://archaeologydataservice.ac.uk

Allen, Kathleen M. S., Stanton W. Green, and Ezra B. W. Zubrow, eds. 1990. *Interpreting Space: GIS and Archaeology*. London and New York: Taylor and Francis.

Antiquity À-la-carte. n.d. Antiquity À-la-carte. Ancient World Mapping Center. Accessed January 13, 2020. http://awmc.unc.edu/wordpress/alacarte/

AriadnePlus. n.d. Accessed January 13, 2020. https://ariadne-infrastructure.eu/about-ariadne/

AWMC. 2016. Mapping the Past: GIS Approaches to Ancient History. Ancient World Mapping Center. Accessed January 13, 2020. http://awmc.unc.edu/wordpress/conference-2016-mapping-the-past/

Blevins, Cameron. 2014. "Mining and Mapping the Production of Space. A View of the World from Houston." Accessed January 13, 2020. https://web.stanford.edu/group/spatialhistory/cgi-bin/site/pub.php?id=93

Bodenhamer, David J., John Corrigan, and Trevor M. Harris, eds. 2010. *The Spatial Humanities: GIS and the Future of Humanities Scholarship*. Bloomington, IN: Indiana University Press.

Bodenhamer, David J., John Corrigan, and Trevor M. Harris, eds. 2015. *Deep Maps and Spatial Narratives*. Bloomington, IN: Indiana University Press.

Braudel, Fernand. 1949. *La Méditerranée et le monde méditerranéen a l'époque de Philippe II*. Paris: Colin.

Center for Open Science. n.d. "Policy and Culture Change at COS: A Strategic Overview." Accessed January 13, 2020. https://cos.io/policy/

CHGIS. n.d. China Historical Geographic Information System. Accessed January 13, 2020. https://sites.fas.harvard.edu/~chgis/pages/intro/

Cosgrove, Dennis E. 1984. *Social Formation and Symbolic Landscape*. London: Croom Helm.

Cowgill, George L. 1967. "Computer Applications in Archaeology." *Computers and the Humanities* 2 (1), 17–23.

DAAHL. n.d. Digital Archaeological Atlas of the Holy Land. Accessed January 13, 2020. https://daahl.ucsd.edu/DAAHL/Home.php

DARMC. n.d. The Digital Atlas of Roman and Medieval Civilization. Accessed January 13, 2020. https://darmc.harvard.edu

Dear, Michael, Jim Ketchum, Sarah Luria, and Douglas Richardson, eds. 2011. *GeoHumanities: Art, History, Text at the Edge of Place*. New York: Routledge.

Digital Crete, n.d. Digital Crete: Mediterranean Cultural Itineraries. Accessed January 13, 2020. http://digitalcrete.ims.forth.gr/index.php?l=1

Doran, Jim, and Frank R. Hodson. 1975. *Mathematics and Computers in Archaeology*. Edinburgh: Edinburgh University Press.

Döring, Jörg, and Tristan Thielmann, eds. 2008. *Spatial Turn: das Raumparadigma in den Kultur- und Sozialwissenschaften*. Bielefeld: Transcript Verlag.

EarthWorks. n.d. Stanford Libraries: EarthWorks, Scanned Maps. Accessed January 13, 2020. https://earthworks.stanford.edu/catalog?featured=scanned_maps&q=

Elliott, Thomas, Sebastian Heath, and John Muccigrosso, eds. 2014. *Current Practice in Linked Open Data for the Ancient World. ISAW Papers* 7. New York: Institute for the Study of the Ancient World. doi.org/2333.1/gxd256w7

Ethington, Philip J. 2007. "Placing the Past: 'Groundwork' for a Spatial Theory of History." *Rethinking History: The Journal of Theory and Practice* 11 (4), 465–93.

Faris, Jack, Evelyne Kolker, Alex Szalay, Leon Bradlow, Ewa Deelman, Wu Feng, Judy Qiu, Donna Russell, Elizabeth Stewart, and Eugene Kolker. 2011. "Communication and Data-Intensive Science in the Beginning of the 21st Century." *Omics: A Journal of Integrative Biology* 15 (4), 213–15. https://doi.org/10.1089/omi.2011.0008

Fiore-Gartland, Brittany, and Anissa Tanweer. 2019. "Community-Level Data Science and its Spheres of Influence: Beyond Novelty Squared." University of Washington e-Science Institute. Accessed January 13, 2020. https://escience.washington.edu/community-level-data-science-and-its-spheres-of-influence-beyond-novelty-squared/

Gaffney, Vincent, and Zoran Stančič. (1991). *GIS Approaches to Regional Analysis: A Case Study of the Island of Hvar*. Ljubljana: Znanstveni Instituti Filozofske Fakultete.

Galeazzi, Fabrizio, and Heather Richards-Rissetto. 2018. "Editorial Introduction: Web-based Archaeology and Collaborative Research." *Journal of Field Archaeology*, 43 (sup1), S1–S8. doi.org/10.1080/00934690.2018.1512701

GBHGIS. n.d. Great Britain Historical Geographic Information System. http://www.gbhgis.org

Goodchild, Michael F. "Geographical Information Science." *International Journal of Geographical Information Systems* 6 (1), 31–45. doi.org/10.1080/02693799208901893

Gregory, Ian N. 2014. "Challenges and Opportunities for Digital History." *Frontiers in Digital Humanities* 1, 1–2. doi.org/10.3389/ fdigh.2014.00001

Gregory, Ian N., and Alistair Geddes, eds. 2014. *Toward Spatial Humanities: Historical GIS and Spatial History*. Bloomington, IN: Indiana University Press.

Gregory, Ian N., and Paul S. Ell. 2007. *Historical GIS: Technologies, Methodologies and Scholarship*. Cambridge: Cambridge University Press.

Gregory, Ian N., and Richard G. Healey. 2007. "Historical GIS: Structuring, Mapping, and Analysing Geographies in the Past." *Progress in Human Geography* 31 (5), 638–53. doi.org/10.1177/0309132507081495

Gregory, Ian N., Karen K. Kemp, and Ruth Mostern. 2001. "Geographical Information and Historical Research: Current Progress and Future Directions. *History and Computing* 13, 7–21.

Grossner, Karl, Timothy Hill, Rainer Simon, Richard Light, Arno Bosse, Gabriel Bodard, Mia Ridge, and Wolfgang Schmidle. 2017. "From Linking Places to a Linked Pasts Network." Accessed January 13, 2020. http://linkedpasts.github.io/assets/lp_whitepaper.pdf

Hadjikyriacou, Antonis. 2018a. "Introduction: Ottoman Recogito (OttRec)." 6 June 2018. Accessed January 13, 2020. https://medium.com/pelagios/introduction-ottoman-recogito-ottrec-3157478130fd

Hadjikyriacou, Antonis. 2018b. "Interim Report for the Ottoman Recogito (OttRec) Project." 12 September 2018. https://medium.com/pelagios/interim-report-for-the-ottoman-recogito-ottrec-project-f163ad35cfb

Hadjikyriacou, Antonis. 2019. "Final Report for Ottoman Recogito (OttoRec)." 5 February 2019. https://medium.com/pelagios/final-report-for-ottoman-recogito-ottorec-e85733dd8269

Harrison, Tim. 2018. "Computational Research on the Ancient Near East (CRANE): Large-Scale Data Integration and Analysis in Near Eastern Archaeology." *Levant* 2018, 1–4. doi.org/10.1080/00758914.2018.1492784

Heath, Sebastian. 2018. "SPARQL as a First Step for Querying and Transforming Numismatic Data: Examples from Nomisma.org." In *Alexander the Great. A Linked Open World*, edited by Simon Glenn, Frédérique Duyrat, and Andrew Meadows, 35–52. *Scripta Antiqua* 116. Bordeaux: Ausonius Éditions.

Hrant Dink Foundation. n.d. "Discover Istanbul with KarDes Mobile Application!" Accessed January 13, 2020. https://hrantdink.org/en/activities/projects/cultural-heritage/2177-discover-istanbul-with-kardes-mobile-application

iDAI. n.d. iDAI.world: For Archaeological Research. Accessed on January 13, 2020. https://idai.world

IFEA Map. n.d. IFEA Map Workshop Archives. Accessed January 13, 2020. http://map-archivis.ifea-istanbul.net/s/en/page/welcome

Kansa, Eric C., Sarah W. Kansa, and Ethan Watrall, eds. 2011. *Archaeology 2.0: New Approaches to Communication and Collaboration. Cotsen Digital Archaeology Series*, 1. Los Angeles: Cotsen Institute of Archaeology. http://www.escholarship.org/uc/item/1r6137tb

Kansa, Eric. 2014. "Open Context and Linked Data." In *Current Practice in Linked Open Data for the Ancient World*, edited by Thomas Elliott, Sebastian Heath, and John Muccigrosso. *ISAW Papers* 7. New York: Institute for the Study of the Ancient World. http://dlib.nyu.edu/awdl/isaw/isaw-papers/7/kansa

Kansa, Sarah W., and Eric C. Kansa. 2014. "Data Publishing and Archaeology's Information Ecosystem." *Near Eastern Archaeology* 77, 223–27.

Klein, Holger. n.d. The Istanbul Documentation Project. Accessed January 13, 2020. https://mcid.mcah.columbia.edu/art-atlas/istanbul-documentation-project/about

Knowles, Anne K, ed. 2002. *Past Time, Past Place: GIS for History*. Redlands, CA: ESRI Press.

Ladurie, Emmanuel Le Roy. 1973. *Le territoire de l'historien*. Paris: Gallimard.

Lefebvre, Henri. 1974. *La production de l'espace*. Paris: Éditions Anthropos.

Lock, Gary R. 2003. *Using Computers in Archaeology: Towards Virtual Pasts*. London: Routledge.

Lock, Gary, and John Wilcock. 1987. *Computer Archaeology*. Aylesbury: Shire Publications.

Lock, Gary, and T. M. Harris. 1991. "Integrating Spatial Information in Computerized Sites and Monuments Records: Meeting Archaeological Requirements in the 1990s." In *Computer Applications and Quantitative Methods in Archaeology 1990*, edited by Kris Lockyear and Sebastian P. Q. Rahtz, 165–73. BAR International Series 565. Oxford: BAR.

McPherron, Shannon P., and Harold L. Dibble. 2002. *Using Computers in Archaeology: A Practical Guide*. Boston, MA: McGraw-Hill/Mayfield.

MedArchNet. n.d. Mediterranean Archaeological Network. Accessed January 13, 2020. http://medarchnet.org

MEDMAPS 2005. Mapping Mediterranean Lands. Digital Library for International Research. Accessed January 13, 2020. http://www.dlir.org/medmaps-home.html

Mellaart, James "The Second Millennium Chronology of Beycesultan." *Anatolian Studies* 20, 55–67.

Middell, Matthias, and Katja Naumann. 2010. Global History and the Spatial Turn: from the Impact of Area Studies to the Study of Critical Junctures of Globalization. *Journal of Global History* 5, 149–70. doi.org/10.1017/S1740022809990362

Mitchell, Peta. 2011. "'The Stratified Record Upon Which We Set Our Feet': The Spatial Turn and the Multilayering of History, Geography, and Geology." In *GeoHumanities: Art, History, Text at the Edge of Place*, edited by Michael Dear, Jim Ketchum, Sarah Luria, and Douglas Richardson, 71–83. New York: Routledge.

OCHRE. n.d. The Online Cultural and Historical Research Environment (OCHRE). Accessed January 13, 2020. https://voices.uchicago.edu/ochre/

Open Context. n.d. Accessed January 13, 2020. https://opencontext.org

Paliou, Eleftheria, Undine Lieberwirth, and Silvia Polla, eds. 2014. *Spatial Analysis and Social Spaces: Interdisciplinary Approaches to the Interpretation of Historic and Prehistoric Built Environments*. Berlin: de Gruyter.

Pelagios. n.d. Pelagios Commons. Accessed January 13, 2020. http://commons.pelagios.org.

Peripleo. n.d. Peripleo: the Pelagios Exploration Engine. Accessed January 13, 2020. https://peripleo.pelagios.org

Pleiades. n.d. Accessed January 13, 2020. https://pleiades.stoa.org/.

Recogito. n.d. Recogito. Accessed January 13, 2020. https://recogito.pelagios.org/.

Reilly, Paul, and Sebastian P. Q. Rahtz. 1992. *Archaeology and the Information Age: A Global Perspective*. London / New York: Routledge.

Richards, Julian D. 1998. "Recent Trends in Computer Applications in Archaeology." *Journal of Archaeological Research* 6 (4), 331–82.

Richards, Julian D., and N. S. Ryan. 1985. *Data Processing in Archaeology*. Cambridge: Cambridge University Press.

Scheidel, Walter, and Elijah Meeks. n.d. "ORBIS: The Stanford Geospatial Network Model of the Roman World." Accessed January 13, 2020. http://orbis.stanford.edu/.

Schloen, J. David, and Sandra R. Schloen. 2012. *OCHRE: An Online Cultural and Historical Research Environment*. Winona Lake, IN: Eisenbrauns.

Shanks, Michael, and Christopher Witmore. 2012. "Archaeology 2.0? Review of *Archaeology 2.0: New Approaches to Communication and Collaboration*, edited by Eric C. Kansa, Sarah W. Kansa, and Ethan Watrall. *Internet Archaeology* 32. https://doi.org/10.11141/ia.32.7

Silveira, Luís, E da. 2014. "Geographic Information Systems and Historical Research: An Appraisal." *International Journal of Humanities and Arts Computing* 8 (1), 28–45. doi.org/10.3366/ijhac.2014.0118

Slater, Matthew H., Joanna K. Huxster, and Julia E. Bresticker. 2019. "Understanding and Trusting Science." *Journal for General Philosophy of Science* 50 (2), 247–61. doi.org/10.1007/s10838-019-09447-9

Snow, Catherine E., and Kenne A. Dibner, eds. 2016. *Science Literacy: Concepts, Contexts, and Consequences*. Washington, DC: The National Academies Press.

Spatial Humanities. 2016. Spatial Humanities: Texts, GIS, & Places. Accessed 13 January 2020. https://www.lancaster.ac.uk/fass/projects/spatialhum.wordpress/

Talbert, Richard J. A. 2000. *Barrington Atlas of the Greek and Roman World*. Princeton: Princeton University Press.

TAYGIS. n.d. The Archaeological Settlements of Turkey Geographic Information System. Accessed January 13, 2020. http://www.tayproject.org/giseng.html

tDAR. n.d. The Digital Archaeological Record. Accessed January 13, 2020. https://www.tdar.org

Thomas, Julian. 1993. "The Politics of Vision and the Archaeologies of Landscape." In *Landscape: Politics and Perspectives*, edited by Barbara Bender, 19–48. Oxford: Berg.

Thomas, Julian. 2004. *Archaeology & Modernity*. London: Routledge.

Thompson, Nato, ed. 2008. *Experimental Geography: Radical Approaches to Landscape, Cartography, and Urbanism*. Melville House, New York.

Tilley, Christopher. 1994. *A Phenomenology of Landscape*. London: Berg.

Tuan, Yi-Fu. 1977. *Space and Place: The Perspective of Experience*. Minneapolis: University of Minnesota Press.

Tuzcu, Nil. n.d. Istanbul Urban Database. Accessed January 13, 2020. http://www.istanbulurbandatabase.com

Voorburg, René. 2011. OmnesViae: Roman Routeplanner. Accessed January 13, 2020. https://omnesviae.org/

Voorburg, René. 2014. Vici.org: Archaeological Atlas of Antiquity. Accessed January 13, 2020. https://vici.org/#13/38.61900307285583,27.939359756347688/56646

Warf, Barney, and Santa Arias, eds. 2009. *The Spatial Turn: Interdisciplinary Perspectives*. London: Routledge.

Weber, Martin. n.d. Tabula Peutingeriana. Accessed January 13, 2020. https://www.tabula-peutingeriana.de

Whallon, Robert. 1972. "The Computer in Archaeology: A Critical Survey." *Computers and the Humanities* 7 (1), 29–45. Accessed January 14, 2020. www.jstor.org/stable/30208220.

WHGazetteer. n.d. "World-Historical Gazetteer: A Project of the World History Center at the University of Pittsburgh." Accessed January 13, 2020. http://whgazetteer.org

White, Richard. 2010. "What is Spatial History?" *Stanford University Spatial History Project*. 1 February 2010. Accessed January 13. 2020. https://web.stanford.edu/group/spatialhistory/cgi-bin/site/pub.php?id=29

Wilcock, John D. 1989. "Review Paper: Over Thirty Years' Application of the Computer in Archaeology. *Science and Archaeology* 31, 35–44.

Wilcock, John D., ed. 1973. *Computer Applications in Archaeology* 1. Special issue of *Science and Archaeology* 9. Stafford: George Street Press.

Wright, Holly, and Julian D. Richards. 2018. "Reflections on Collaborative Archaeology and Large-Scale Online Research Infrastructures." Journal of Field Archaeology, 43 (sup1), S60–67. doi.org/10.1080/00934690.2018.1511960

What Would Indy Do? Resisting Post-Truth through the Practice of Annotation

Elton Barker

Entering the Commons

Among the many acts of resistance that welcomed the formal investiture of the 45th president of the United States was self-proclaimed white nationalist Richard Spencer, in Washington to celebrate the inauguration of his president, getting punched in the face. The video of the assault, posted on Twitter, subsequently went viral, as images of famous American heroes were shared with the hashtag #punchaNazi. One of the most popular was a still from Stephen Spielberg's *Indiana Jones and the Raiders of the Lost Ark*, when Indy lands a knockout blow to one of his Nazi adversaries in the back of a truck (Duggan 2017).

Normally mentioning Indiana Jones in the context of archaeological work is enough to prompt weary sighs from colleagues, who are accustomed to having to patiently explain that being an archaeologist is not quite as glamorous as it is depicted in these films. But my purpose for invoking Indy's name here has to do less with affirming (or denying) his archaeological credentials as with his battle against the fascist appropriation of cultural heritage. Along with giving encouragement to white supremacists, Donald Trump's election has ushered in—and, to a certain extent, is clearly a manifestation of—a culture where expertise and informed opinion are to be rigorously denied, far less coveted. Two of the favorite phrases that Trump used during his presidential campaign, and that have subsequently come to define his administration, are "fake news" and "alternative facts." When the challenge to expertise has gone so far that it is not only viewed skeptically but there is an active intolerance of it—when, indeed, the very idea of an epistemological truth or demonstrable fact is denied—then democratic culture is at stake.[1] As experts in our various disciplines, scholars need to

1 Attacks on the sophists, who stand accused in much of the literature that survives from the democratic culture of fifth-century BCE Athens of peddling ways of making the weaker argument the stronger, reflect similar anxieties about the threat to democracy: see the evidence collected in Kerferd 1981.

join the resistance and counter fascist ideology with reasoned argument and empirical observation.[2]

Important here are questions of both infrastructure and agency: we need the web to be as open as possible in terms of its structure and delivery;[3] but *we* also need to be as open as possible in our scholarship, with the ability, at the very least, to be able to point to the evidence on which we base our arguments and assertions.[4] In the language of research communication, our data need to be FAIR: findable, accessible, interoperable, and reusable (Table 1). Or, to put that slightly differently, how can we enable people to discover our work, gain access to it, and (re)use it, particularly when academic publishing regards open access as a threat to its business model?

As researchers who (literally) dig into the past to better understand who we are and where we come from, archaeologists stand on the frontline of the battle. Many indeed have already embraced the challenge of putting their scholarship and research "into the commons."[5] Three indicative examples serve to illustrate the kind of work that is engaged in directly informing the public, as well as moving toward a new kind of scholarship: Open Context provides a process and a service for publishing research data (openly) on the web, preserving those data with leading digital libraries, and providing the means to analyze them through maps and other media (opencontext.org); ODATE, or The Open Digital Archaeology Textbook Environment, is an initiative to facilitate a collaborative practice in publishing, by enabling researchers to contribute their expertise to writing an archaeology textbook within an open publishing framework (o-date.github.io/); the Twitter hashtag #dayofarch invites archaeologists from all around the world to share their experience of an ordinary day working in the field, the results of

2 See, for example, the pushback against the white supremacist appropriation of classical material on social media from some US classicists, in particular: Bond 2017, Futo-Kennedy 2017, and Zuckerberg 2017, all of whom have been predictably trolled for being women or liberals or both. The UK has not been immune from such controversies either, as evidenced by the debate over a BBC school video that featured a family in Roman Britain in which the father, a high-ranking soldier, was depicted as non-white (Beard 2017). On accountability and transparency as fundamental procedures of fifth-century BCE Athens, see Vernant ([1962] 1982, 51–52).

3 As the debates around net neutrality have made clear: e.g., https://www.freepress.net/issues/free-open-internet/net-neutrality.

4 In relation to the "painted statues" and "ethnic Roman Britain" stories, Neville Morley (2017) suggests that "what is surprising about these two arguments is that the substantive issues—ancient statues were painted, the Roman Empire (including Britain) was ethnically diverse—are such old hat." What is needed, according to Morley, is for academics to help people move beyond the myths and half-truths they may have acquired at school or from the media, so that they can better "understand how historians (and scientists) actually go about investigating and reconstructing the past."

5 Marcel Detienne (1996, 95) explains the expression "into the middle" (*es meson*) as to set things in common, whether the referent is material goods (as when Achilles puts "into the middle" booty to be competed over in Homer's *Iliad*, e.g., 23.507), or arguments and opinions (in political debate). Leslie Kurke (1999, 29) similarly describes Herodotus's *Histories* as an "open *agora* [or market-place] of *logoi* [accounts]," which allow different voices to be heard; cf. Dewald 1987. On the use of the expression "*es meson*" in representations of debate in classical Greek literature: Barker 2009, 17–18 and the Index, s.v. "*meson, es*."

Table 1. FAIR principles. Source: Adopted from FORCE11 (2017).

Concept	Principle
To be Findable	F1. (Meta)data are assigned a globally unique and eternally persistent identifier.
	F2. Data are described with rich metadata.
	F3. (Meta)data are registered or indexed in a searchable resource.
	F4. Metadata specify the data identifier.
To be Accessible	A1. (Meta)data are retrievable by their identifier using a standardized communications protocol.
	A1.1. The protocol is open, free, and universally implementable.
	A1.2. The protocol allows for an authentication and authorization procedure, where necessary.
	A2. Metadata are accessible, even when the data are no longer available.
To be Interoperable	I1. (Meta)data use a formal, accessible, shared, and broadly applicable language for knowledge representation.
	I2. (Meta)data use vocabularies that follow FAIR principles.
	I3. (Meta)data include qualified references to other (meta)data.
To be Re-Usable	R1. Meta(data) have a plurality of accurate and relevant attributes.
	R1.1. (Meta)data are released with a clear and accessible data usage license.
	R1.2. (Meta)data are associated with their provenance.
	R1.3. (Meta)data meet domain-relevant community standards.

which are then brought together for hosting on a common website (dayofarchaeology.com). Such examples could be multiplied; the point is that archaeologists are leading the way in the effort to do their research in the public domain.

Within this context of enabling research methodologies and findings to be heard in the community's "contest of voices" (Goldhill 1986, on fifth-century BCE Athenian tragedy), I want to focus on one small, but critical, element of digital practice: semantic annotation. The idea of annotation—of adding notes to a document—goes back at least as far as medieval manuscripts, out of which the vast majority of our classical corpus of Greek and Latin literature is reconstructed (Barker and Terras 2016). Figure 1, for example, shows the first page of the Venetus A manuscript of Homer's *Iliad*, complete with scholarly annotations both in the margins and inline. Such annotations provide further information about a particular idea or element in the text to enrich its interpretation, often by pointing to information external to the text itself. Once frowned upon as a practice by the custodians of a printed book culture (viz., librarians), the annotation is making something of a comeback in the digital medium, whereby metadata (some

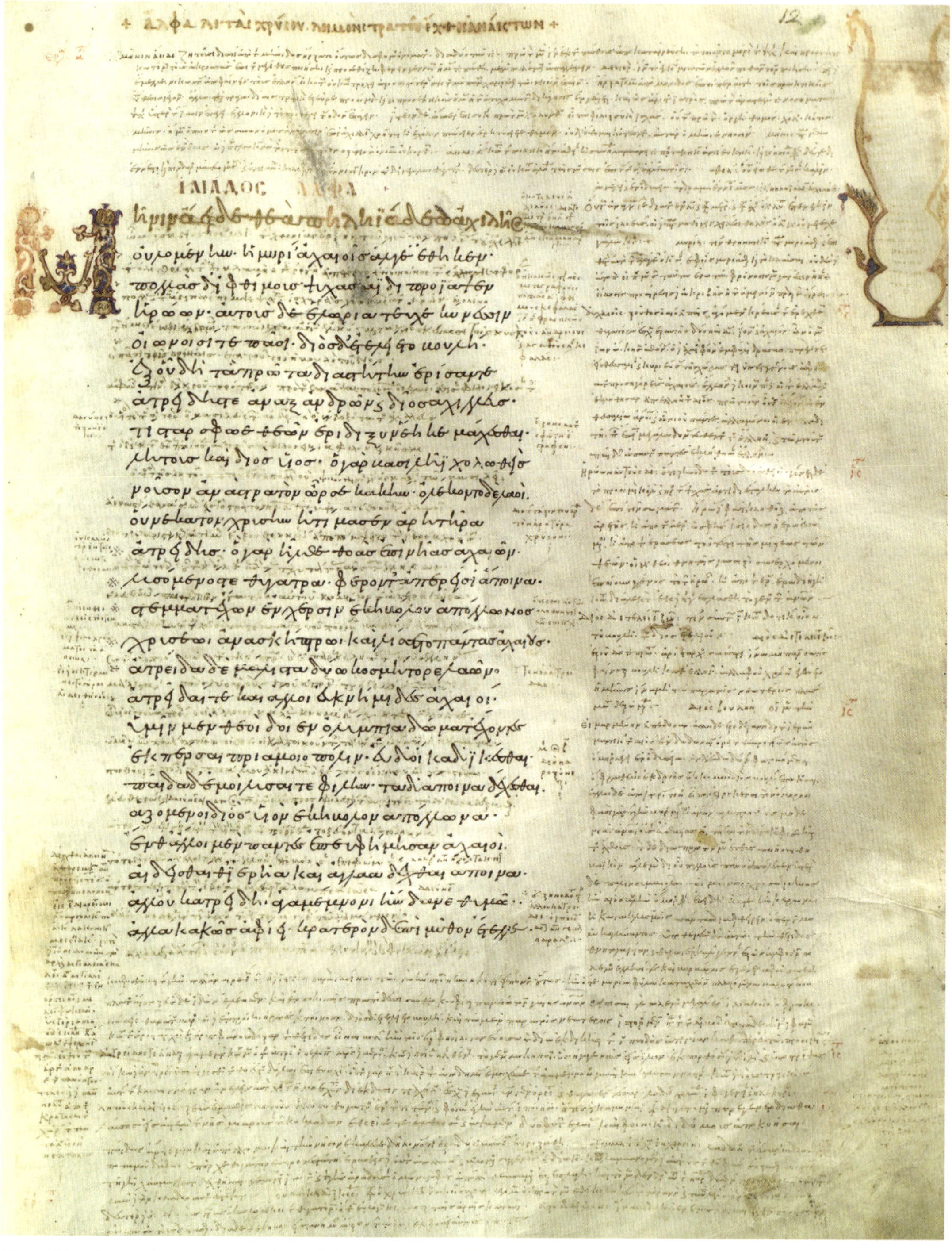

Fig. 1. Marcianus Graecus Z.454 (= 822) (Venetus A), folio 12 recto. Homer *Iliad* Book 1, lines 1–1.25. http://www.homermultitext.org/. urn:cite:hmt:vaimg.VA012RN-0013. This image was derived from an original (©2007, Biblioteca Nazionale Marciana, Venice, Italy). The derivative image is ©2010, Center for Hellenic Studies. Original and derivative are licensed under the Creative Commons Attribution-Noncommercial-Share Alike 3.0 License. The CHS/Marciana Imaging Project was directed by David Jacobs of the British Library.

kind of comment, explanation, or presentational "markup") are attached to the online document to enhance its reading. Such information might relate to additional information other than the words on the page, including author, title, provenance, and date, or section demarcations, like book, chapter, paragraph (see the discussion below on TEI: a particular form of XML markup). Metadata can also relate to entities within the text, too, such as the people, places, or events mentioned. Information of this kind can give a text added value, by which the "hard copy" text (whether manuscript or book) is supplemented with additional semantic information.

For the rest of this chapter I want to reflect on the value of digital annotation through my own experience on two projects, in which we have used the concept of place as a starting point for analyzing historical documents.

Spatializing the Humanities

The idea of place and space is profoundly important in humanist thought, be it the spatial organization of narrative, the history of places, or location-oriented analysis of cultural objects. Space is not simply an empty container for historical action but a significant medium for the development and understanding of culture. It occurs, writes Michel de Certeau (1984, 117–18), "as the effect produced by the operations that orient it, situate it, temporalize it, and make it function as a polyvalent unity of conflictual programs or contractual proximities." People are born, live, and die in places; events always happen somewhere. According to eminent geographer Yi-Fu Tuan (1977, 6), "What begins as undifferentiated space becomes place as we get to know it better and endow it with value."

Over the past decade the extraordinary growth of new technologies has reinvigorated the humanities' spatial turn and given rise to a new discipline: the spatial humanities (Dear, Ketchum, and Richardson 2011). Dominating this field of largely historical-based research has been Geographic Information Systems (GIS). As the name implies, a GIS is a system designed to capture, store, manipulate, analyze, manage, and present spatial or geographic data. Mapping information can have a surprisingly powerful effect. As David Bodenhamer (2010, 28) suggests, simply by virtue of "locating historical and cultural exegesis more explicitly in space and time . . . [GIS] finds patterns, facilitates comparisons, enhances perspectives, and illustrates data." Ian Gregory and David Cooper (2010, 89) discuss the capacity of GIS to enable scholars to explore "subjective geographies through the spatial representation of qualitative, or fuzzy, data." For research into the ancient world, GIS provides the technical backbone to and interface for two important, open-web resources: *Antiquity À-la-Carte* (awmc.unc.edu/wordpress/alacarte/), which allows researchers to map and analyze historical data produced by the Ancient World Mapping Center; and ORBIS (orbis.stanford.edu/), which enables queries of the time and financial cost associated with different types of travel in antiquity.

Yet, even as such digital mapping technologies are enabling the emergence of the spatial humanities, critical geographers have been questioning the objectivity of the cartographic representations on which these technologies tend to rely. Conventional maps, in the words of David Harvey (1996, 4), are "typically totalizing, usually two-dimensional, Cartesian, and very undialectical devices." Indeed, there has been a growing awareness that a history of space (and place) should also include "the history of representations" (Lefebvre 1991, 42), or that the spatial information embedded in non-cartographic material (like texts) have the potential to un-think the Cartesian tradition that has dominated understandings of space since the Enlightenment (Gurevich 1985). In the Greco-Roman world, maps appear to have been relatively rare commodities; even when they are mentioned in our sources, they are treated with suspicion or flatly criticized. In Aristophanes's satire on the latest intellectual thinking, *Clouds* (423 BCE), for example, Strepsiades, who represents in some way the Athenian common man, is shown a map by a student of "Socrates": as Sparta, Athens's great enemy, is pointed out to him, he leaps back and exclaims, "So close to us? How about moving it a lot farther away from us!" (*Clouds* 215; ὡς ἐγγὺς ἡμῶν. τοῦτο πάνυ φροντίζετε, / ταύτην ἀφ᾽ ἡμῶν ἀπαγαγεῖν πόρρω πάνυ). It is a joke that plays on the fantasy that the map accurately embodies the space that it depicts (cf. Harley 1989). This is a rare example, however: Greco-Roman spatial knowledge is primarily mediated through poetry, drama, or historiographical narratives. These kinds of *discursive* representations present a challenge to traditional GIS, with its positivist emphasis on precise measurement and categorization. It is not only the case that GIS is notoriously poor at handling uncertain, incomplete, and ambiguous data (Gregory and Healey 2007), but also that *text*-based research studies of humanistic enquiry "raise fundamental epistemological and ontological issues for GIS applications" (Harris, Bergeron, and Rouse 2011, 228). Other ways of capturing and reimagining textual space are needed.

The Forms of Place and the Spatial Form of the *Histories* of Herodotus

Roughly contemporary with *Clouds* is Herodotus's *Histories*, a rich example that I would like to pick through with reference to a project I directed: Hestia (hestia.open.ac.uk; see, for example, Barker et al. 2010; 2016). In this initiative we were interested in identifying and exploring two aspects of textual space: first, what places does Herodotus mention in his text and how does he describe them (the forms of place); and second, how his narrative is underpinned by and, to a certain extent, structures or organizes a sense of space (its spatial form). At the center of the *Histories* lies a similar trope that has just been seen in the example of Aristophanes, as Aristagoras, the tyrant of Miletus, arrives in Sparta seeking support for a revolt from Persian rule, armed with "a bronze picture, on which the whole world was engraved and all the seas and all the rivers" (ἔχων χάλκεον πίνακα ἐν τῷ γῆς ἁπάσης περίοδος ἐνετέτμητο καὶ θάλασσά τε πᾶσα καὶ ποταμοὶ πάντες, 5.49.1). While Aristagoras exploits the representational form of this "map" to promote

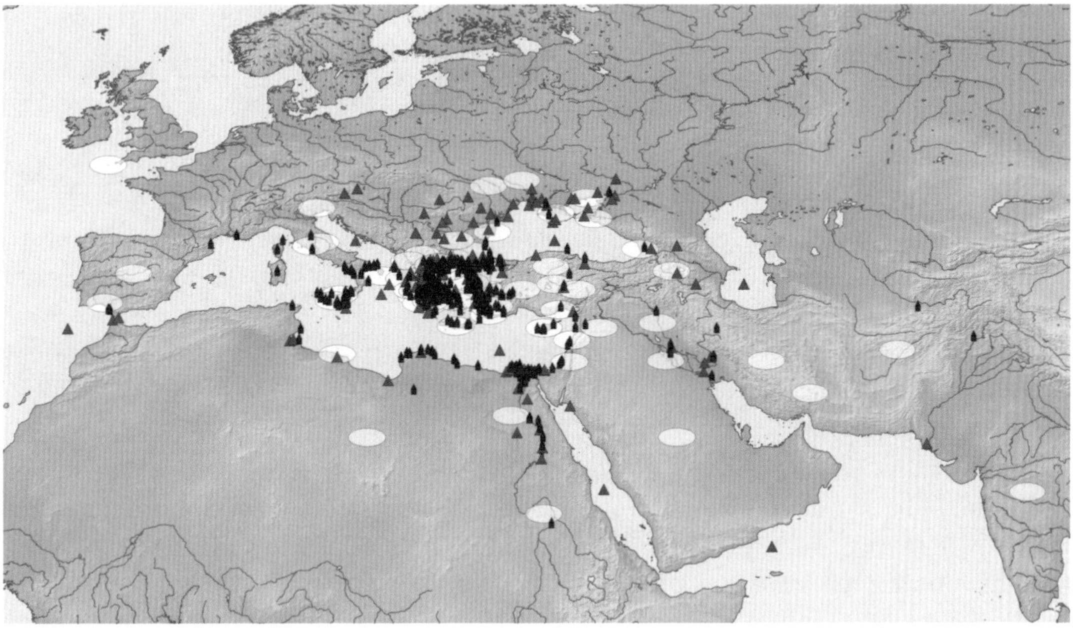

Fig. 2. Map (in QGIS) showing all places mentioned in Herodotus's *Histories* according to settlement (house symbol), territory (oval), and physical feature (triangle).

the idea of easy conquest, Herodotus juxtaposes to this episode his own discursive representation of the same territorial scope (5.50–52), in a way that invites his reader to reflect on space as something lived and experienced and far more complex. Elsewhere, Herodotus laughs at the maps produced by his Ionian contemporaries that divide the world into two regions of equal size (4.36.2) and is also careful to put the claim that Asia belongs to the Persians and Europe to the Greeks in the mouth of certain Persian writers (1.4.4). It seemed to us working on Hestia that Herodotus's representation of space was worthwhile exploring *precisely because* it was discursive and, in some shape or form, *counter*-cartographic (Purves 2010)—an anti-GIS, if you will.

Our approach was intentionally experimental and as innovative as possible, particular in our application and use of digital resources and mapping software, with varying degrees of success (Barker, Isaksen, and Ogden 2016); but at the heart of our investigations was the idea of semantic annotation. As a start, we reused a digital text of the *Histories* that was freely available from the Perseus Digital Library (via a CC-BY license). Encoded according to a particular XML annotation schema designed for textual analysis (TEI: tei-c.org), this text came with most of the places in the *Histories* already marked up. As such, it was easy to extract the place information and plot it in a GIS, which is what we did (Fig. 2). Even as a simple data cleaning exercise, this visualization was extremely useful, since we could easily spot "outliers" and correct their misattributions (such as "Priam"—the king of Troy—being incorrectly marked as the town in the United States). But it was also useful for revealing at a glance two important ideas for interpreting Herodotus's spatial representation: first, that, while the spatial scope of Herodotus's narrative was broad, its focus was concentrated on the area of the Aegean; and, second, that the farther one moved from this center, the more likely physical features were used in description (rather than places). On the other hand, the representation of Herodotean places as "dots on a GIS map" simply served to reproduce the exact same Cartesian worldview that we had been at pains to challenge. Indeed, such a GIS visualization of Herodotus presented the danger not only of reproducing Cartesian principles but even of aestheticizing them (Wood 2010). Our thesis was that textual space was constructed by and through the stories told about it; point-based geometry was simply insufficient to capture this complexity, fluidity, plurality, ambiguity, and interconnectedness.

We came back, then, to the question of how Herodotus relates the places mentioned in his narrative: how, in other words, the narrative form *constructs* a sense of space. In taking this approach we were drawing on earlier work that had conceived of Herodotus's representation of space as *hodos*, a "way" or "path" (or *hodological*: Purves 2010, 15; cf. Janni 1984), as well as more broadly the notion of space (or *topos*) as conveying "movement" (*kinesis*; sic. *topokinesis*: Turnbull 2007). This idea of movement (through space or of places) was inspired by our reading of Herodotus's description of his narrative journey: "I'll go forward with the further part of my account, going through small and great cities alike. For those that were once great have now become small, while those that were great in my time were before small" (*Histories* 1.5.3–4).

Subsequently, we found Franco Moretti's (1998, 5) work on literary maps helped to provide a useful guide for bringing out the "place-bound nature of literary forms" and the internal (spatial) "logic of narrative." I would now add to this Caroline Levine's (2015, 16) recent emphasis on tracking "the forms of the content, the many organizing principles that encounter one another inside as well as outside of the literary text . . . structuring and patterning experience . . . but also competing and colliding and rerouting one another" for drawing attention to the possible political implication underpinning Herodotus's spatial mapping.

Within this more self-consciously theoretical framework, we annotated the text ourselves, by hand (see Bouzarovski and Barker 2016). Our aim was not only to identify spatial entities in a more nuanced and comprehensive manner than what had been possible by relying solely on the data capture from the digital text; above all, we also wanted to mark the relations between these spatial entities in a semantically meaningful way—that is, at the sentence level, where one entity (or more) was the subject of a verb and one (or more) the predicate. After a close reading of one book—Book 5 of the *Histories* was chosen on the basis that it stands at the center of the narrative and represents the coming together of the Greek and Persian worlds—we decided on two kinds of annotations that best reflected our aims. First, in addition to places we marked "proxies"—the individuals, social groups or even non-human agents that belong to the geographical imagination described by Herodotus; place-names alone were insufficient to capture the full complexity of spatial entities depicted in Herodotus's *Histories*. Second, we identified the *textual relatedness* of the places and/or proxies according to their directionality (which place or proxy was the subject and which the predicate) and semantic association: that is, whether the verb joining them denoted movement (*hodology*) and/or transformation (*topokinesis*). Based on these qualities, we determined four kinds of classification:

1. Positioning (including verbs of inclusion, proximity, comparison, origin)
2. Movement (in, through, away from)
3. Intervention in place (e.g., governance, alliance, awareness, commemoration)
4. Intervention with movement (e.g., invasion, exchange, communication, settlement)

Sample diagrams depicting the connections that we drew up are given in Figures 3–5.

The first thing to note about these graphs is their density and complexity. They are difficult to read. And they are difficult to read in two senses. On the one hand, we were faced with the challenge of simply enabling the data to be viewed more easily: Figures 3–5 represent the same graphs (the "total network" of the *Histories* Book 5), but using different software or formatting choices to render their information more readable (essentially by making decisions what to highlight and what to leave out). On the other, we also needed to render the visualizations in such a way that could help inform our analysis. Although the product of intense and laborious line-by-line analysis of an entire book, the graphs do not depict "results" in any conventional understanding of the term. They demand interpretation.

Both aspects of the (problematic) readability of network visualizations have been raised by historian Irad Malkin. A pioneer of network theory to rethink ancient Mediterranean space (and reconnect Greek peoples to their far-flung multicultural places), Malkin is highly skeptical of the utility of graphic visualizations of Mediterranean networks. Such "two-dimensional representations of connectivity," he complains, "mostly turn out to be messy 'spaghetti monsters,'" which not only need "very long verbal explanations" but also omit entirely "the networks' multidirectionality, multidimensionality, and multitemporality" (2011, 18). In short, "It is not enough to draw a few points on the map, identify them as 'Phokaian,' and come up with a diagram of a Phokaian network" (2011, 149). As we discovered, even once we had slayed the "spaghetti monsters," our visualization was still difficult to read. Generally, humanities scholars lack the visual literacy to appreciate, and analyze, the complexity and nuance of graphic data; moreover, were our graph self-explanatory, it would still lack any of the metrics or rules that help enable network analysis. Rather, our graphs are more like the baby-steps of Moretti's network picture of Shakespeare's *Hamlet* than

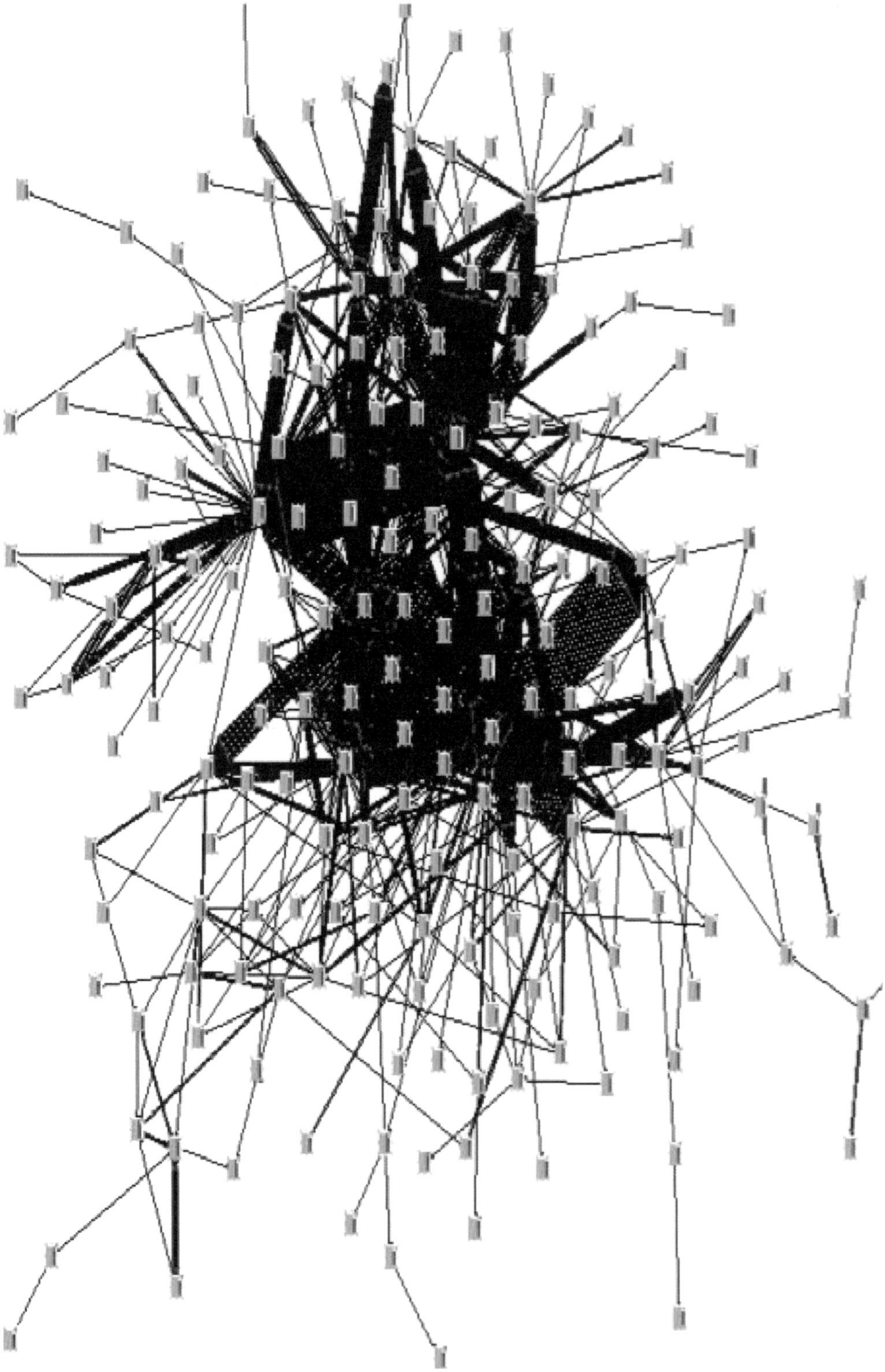

Fig. 3. The total graph network of place (and people) relations in Book 5 of Herodotus's *Histories*, visualized in Gephi, and demonstrating difficult readability.

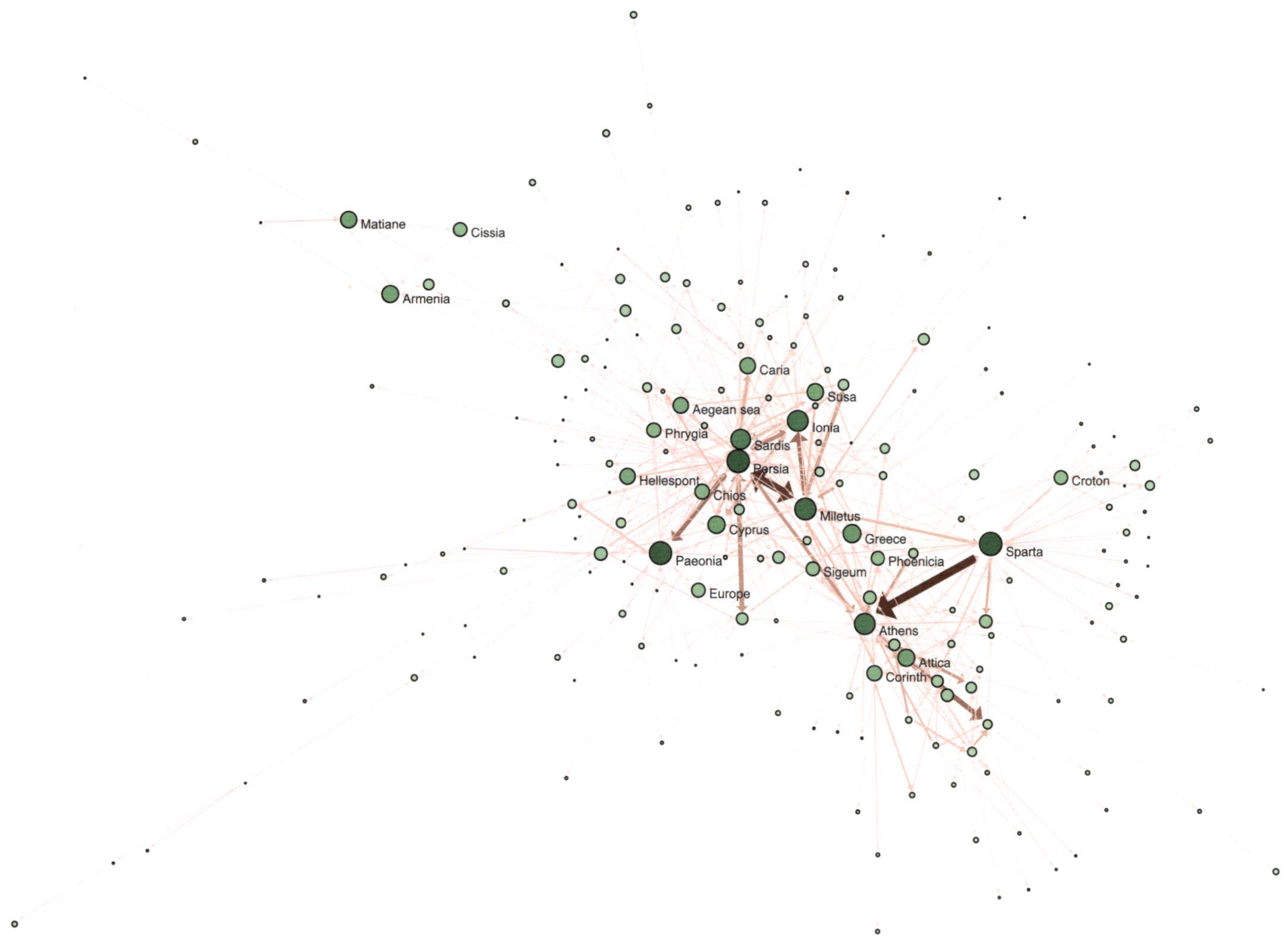

Fig. 4. The total graph network of place (and people) relations in Book 5 of Herodotus's *Histories*, visualized using the Science of Science (Sci2) Tool. Indiana University & SciTech Strategies (http://sci2.cns.iu.edu), with the support of Scott Weingart. The size of the node shows the strength of the relations of each place: the bigger the node, the more relations a place enjoys.

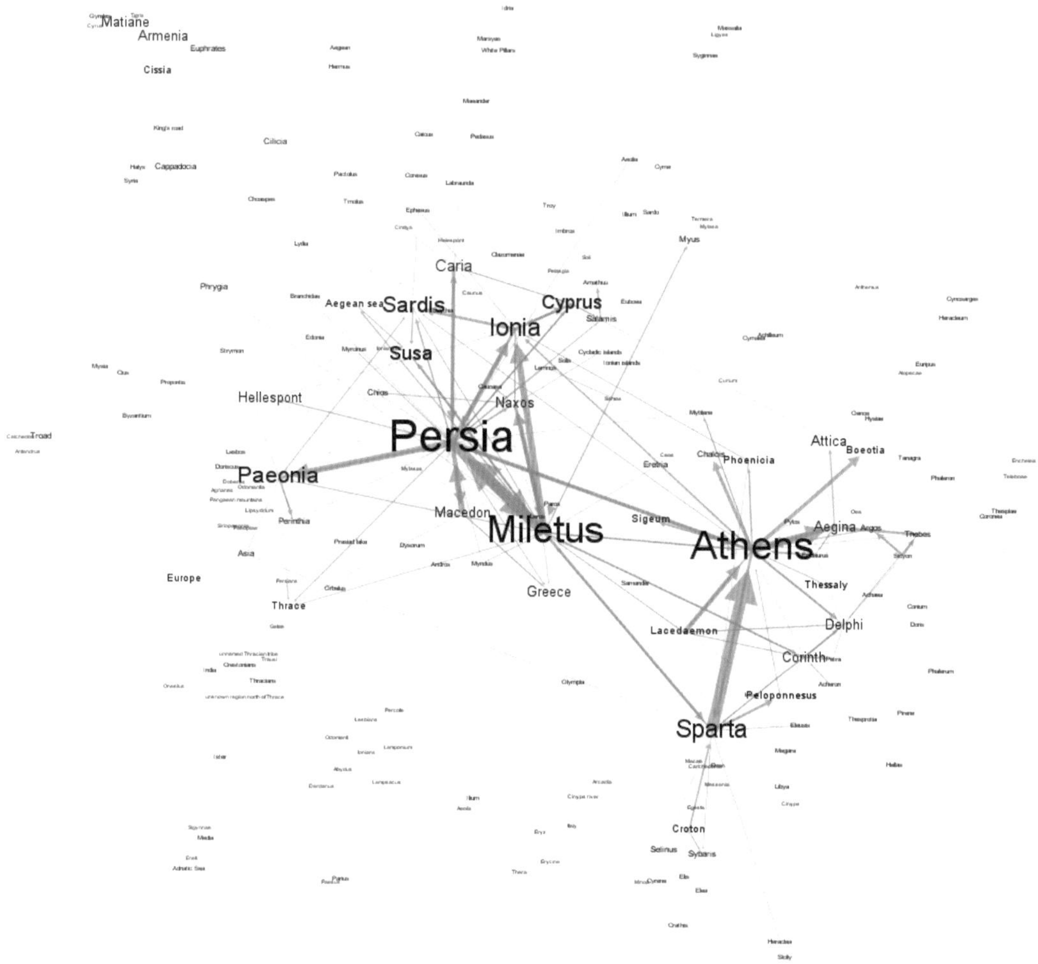

Fig. 5. The total graph network of place (and people) relations in Book 5 of Herodotus's *Histories*, visualized using the Science of Science (Sci2) Tool. Indiana University & SciTech Strategies (http://sci2.cns.iu.edu), with the support of Scott Weingart. The size of font shows the strength of the relations of each place: the bigger the font, the more relations a place enjoys.

those used in the science of network theory (Moretti 2011). In fact—and this is our main point—our graphs are better understood as *provocations* rather than illustrations, as stimulations for further analysis and reflection. They are tools for argument, not the proof or replacement of it.

It is as provocations, I contend, that our graphs can have a strong analytical value by challenging dominant Cartesian-based assumptions of Herodotean space. Such "X-ray" maps (Moretti 2011, 4) bring to the fore the underlying ways in which Herodotus constructs space in his *Histories* in terms of relations between peoples and places rather than according to topographic reality. In Figure 5, which represents our best current attempt at representing the "total network" graph for Book 5, those places (or peoples) most connected are depicted at the center of the graph; the bigger the font, the more times the place is mentioned in relation to others; arrows show the direction of the relation, whether one place is acting on, or being acted on, by another. What this graph reveals, we argue, is a picture of two emerging spheres: an Eastern world centered on and around Persia, and a Western, Greek one with Athens and Sparta dominant. Such a division not only refers back to, and picks up on, Herodotus's own initial distinction between Greeks and barbarians (1.1.1); it is one familiar to us, as implied by the use of the cardinal descriptors "Eastern" and "Western," which come with so much ideological baggage (Said 1978). Yet, the picture is in fact far more complex than a simple geographically based ideological distinction between East and West (cf. Pelling 1997). In addition to the two spheres interlocking in various ways, a number of Greek places—from islands such as Cyprus, Naxos, and Chios, to broader regions such as Ionia, the Aegean Sea, Macedonia, and even the notion of "Greece" itself—have become attracted into the Persian sphere. This picture not only represents the growing power and influence of this super-state over the course of Book 5; it also suggests the alterity of these Greek places, caught between the Greek mainland proper and the fringes of the Persian Empire, and caught up in the growing opposition between the two. In fact, it is the Greek city of Miletus that acts as the fulcrum around which the Greek and Persian worlds pivot and collide: the even bidirectionality of the arrow between Miletus and Persia shows that agency flows both ways in this relationship and reminds us of the activity of Aristagoras, tyrant of Miletus, in engineering conflict against the Persians. The Greek sphere is even less homogenous. Characterized by strong internal dynamics, this sphere reveals the strongest relationship in the whole graph: Sparta's actions on (and against) Athens. This history (and not only at this point) is also about the internal conflicts among Greek cities and peoples, in ways that both anticipate and presage the outbreak of the disastrous war of Herodotus's time (Barker and Pelling 2016).

Such diagrammatic forms, then, offer alternative pathways through Herodotus's space that subvert our long-standing cartographic tradition and suggest new ways of thinking about pre-modern narrative spatial thinking. They present, among other things, (1) ways of thinking about how Herodotus constructs space (or, rather, how events in his *Histories* are embedded in and structured by certain places); (2) ways of contesting essentialist views of space ("Persia," "Athens," etc.), by refocusing attention on the relatedness, intersections, and movement of places and peoples; and (3) ways of identifying and analyzing how space and place are constructed by historical agents (Pred 1983). Fundamentally, such graphs need to be viewed not as illustrations of the data so much as provocations to analyze them further, precisely to elicit the "long verbal explanations" of which Malkin (2011, 18) complains. Depicting network graphs means little without understanding how the networks operate, in particular how these links "jeopardize, stabilize, or reroute bounded unities" (to again quote Levine 2015, 120), or produce, in the words of Doreen Massey (2005, 158), "constellations of trajectories which, though interacting and undoubtedly affecting each other, have very different rhythms." Thinking of maps as provocations, however, presents a further challenge to their visualization of data. As Irad Malkin (2011, 18) has asserted, "We are in need of new modes of graphic representation that would avoid the more precise, yet messy, 'lines connecting dots' and avoid the pitfalls of dazzling oversimplifications." Missing, for example, from our series of network graphs is the concept of time, whether we mean chronological "real" time or narrative time, the unfolding of the narrative or chronotope (Bakhtin 1981). Yet, since relations do not remain immutable but change according to historical circumstances (even over the course of a single book), they need to be located in time as well as space.

What we are talking about here is the need for a new kind of map that better enables exploration and analysis on the part of the viewer, or, as David Bodenhamer (2015, 23) suggests, is "alert to the basic ways in which we analyze space—movement, direction, proximity, and connection, among others—and discover the patterns that occur within and among places." Such a map would be a *deep* map. The deep map has been recently defined as "afford[ing] open-ended exploration of a particular place" by being "more fully situated archives in which one might find myriad traces of evidence about a site, and from these begin to build stories and arguments" that "would represent the pathways that weave places together, creating urban life by their interplay and motion" (Ridge, Lafreniere, and Nesbit 2013, 177–178). It combines "highly structured, mathematically precise cartographic exercises with more subjective and discursive mapping methods" in ways that defamiliarize topographic space to reveal "a multi-layered network of connections between people and places, past and present" (Gregory-Guider 2004, 1–2). Thus deep maps are emerging as a concept for managing heterogeneous data because of advances in technology, by virtue of which they can be "inherently unstable, continually unfolding and changing in response to new data, new perspectives, and new insights"—in essence, "the means by which we represent the contested meanings of space and place, as well as the dynamics that produce them" (Bodenhamer 2015, 21).

Linking Information through Common References to Places

The challenges of visualization still need to be addressed, though progress is being made particularly in the design of user-led exploration of networks (e.g., Dörk, Comber, and Dade-Robertson 2014). For the final part of this chapter, however, I want to return to the idea of semantic annotation and its powerful use as a means of linking the kinds of heterogeneous data I've just described as potentially forming the basis of a deep map. Thus far we have explored place in one document, the narrative history of Herodotus of Halicarnassus. Yet places have a much richer history than can ever be captured in any one representation, cartographic or otherwise. Places "gather things in their midst"—both animate and inanimate entities, as well as experiences and histories. “The process of gathering (things, emotions, people, memories, etc.),” according to Edward Casey (1996, 24, quoted in Cresswell 2015, 51), “suggests that there is a relationship between the inside of a place (which gathers) and an outside (from where things are gathered). It underlines the relational nature of place—the necessity of place being related to its outside.” Or, following Doreen Massey (2005, 130) again: “One way of seeing ‘places’ is as on the surface of maps . . . But to escape from an imagination of space as a surface is to abandon also that view of place. If space is rather a simultaneity of stories-so-far, then places are collections of those stories, articulations within the wider power-geometries of space. Their character will be a product of these intersections within that wider setting, and of what is made of them. And, too, of the non-meetings-up, the disconnections and the relations not established, the exclusions. All this contributes to the specificity of place.” Both critics point to the importance of reading places as fundamentally *relational*, as both connected to other external entities and linked in within wider power-geometries of space.

One opportunity that is emerging for this kind of analysis is through the method of linking data through semantic annotation. By means of example, and by way of returning us to the ideas about openness, transparency, and accountability with which I began, I end this chapter by discussing a second initiative that I have been helping to direct: Pelagios.

Using the linking potential of the web, Pelagios has been pioneering a means of “mutual contextualization,” whereby any online record—be it a text, image, or database—can be connected to another simply by virtue of having something in common with it. The Pelagios method is to achieve linking through the semantic annotation of place references in web documents. That is to say, whereas other mechanisms for linking data require agreement at the level of ontology (essentially a standard data representation) or seek to aggregate data all in one place, Pelagios works through a decentralized modus operandi that uses the method of semantic annotation to enrich metadata in the form of “stand-off” markup.

The method of annotation is twofold (see Simon et al. 2017). First, locate the place(s) in an online document and create an annotation; then align or “resolve” these annotations to a digital authority that provides the means to identify and disambiguate between different places. Digital authority files for places are online gazetteers: as well as providing information about places that are useful for their understanding—items such as geographical coordinates, type of place, political affiliation, bibliography, etc.—online gazetteers also provide what are called uniform resource identifiers (URIs). These URIs essentially operate like social security numbers to help disambiguate places of the same name or identify the same place with different names or typography (Simon et al. 2016a). By annotating the places in a document with URIs, not only can they be correctly identified and placed on a map, if desired, but also, because other data providers are also using the same identifiers, disparate documents can be linked through their references to common places. By these means, Pelagios is able to link two different kinds of knowledge communities—those who hold information about places (the gazetteer community) and those whose documents refer to those places (the research community that depends on place identification and disambiguation).

Originally a trial collaboration to interlink various online data providers who had the technical know-how to align datasets, Pelagios now offers any researcher, data curator, student, or general enthusiast a simple, intuitive means to encode place information in their work, whether or not they are curating data or are interested in producing it as linked data. This do-it-yourself, open-access, and open-source (geo)annotation platform is called Recogito.[6] Thus, where in the Hestia project we had to annotate place and space information by hand, essentially by inputting the information into a spreadsheet (a CSV file), any user is now able to annotate the text itself directly, and a CSV file is only one of a number of output options for the data. In addition, Recogito lowers the barrier of entry for annotating a document. Figure 6 shows the Recogito page for uploading a document for annotation, which can be a text (plain text or TEI encoded), an image (e.g., a map), or a database. Figure 7 shows the annotation screen, where those entities in the document to be marked as places are identified and then aligned or resolved to the appropriate entry in the appropriate gazetteer.[7] Figure 8 shows the map view in which all the newly annotated places are presented in a map interface, including the option of displaying "snippets" of text (if appropriate). This built-in visualization means that anyone can now identify and plot places in a narrative to reveal their spatial footprint and create maps out of texts. Of far greater significance, however, Recogito also provides other download options—KML, RDF, GeoJSON, and CSV—for use in other applications, such as GIS or network analysis tools, as well as for publishing the annotations as linked data. It should also be noted that Recogito provides full capability to annotate image documents (such as pre-modern maps), with tools including zoom, rotate, and transcribe, all with IIIF (International Image Interoperability Framework) support.

What I have been describing is very much the technical structure of Recogito in terms of the options that it provides for annotating and then making use of the annotations. But there is an important social element too. Due to community demand, development of Recogito has evolved to enable social annotation, by which means one can share documents with others and work on the same document synchronously, which has a significant role to play both for collaboration and in the classroom. Furthermore,

6 To access Recogito, simply log in to recogito.pelagios.org/. For all the (open-source) code produced by Pelagios, see github.com/pelagios. For more information about Pelagios and its community at the time of going to press, see pelagios.org and medium.com/pelagios.

7 This two-step process—identifying the place (target) and resolving it to the gazetteer (body)—is essentially producing RDF triples, the data structure for linked data.

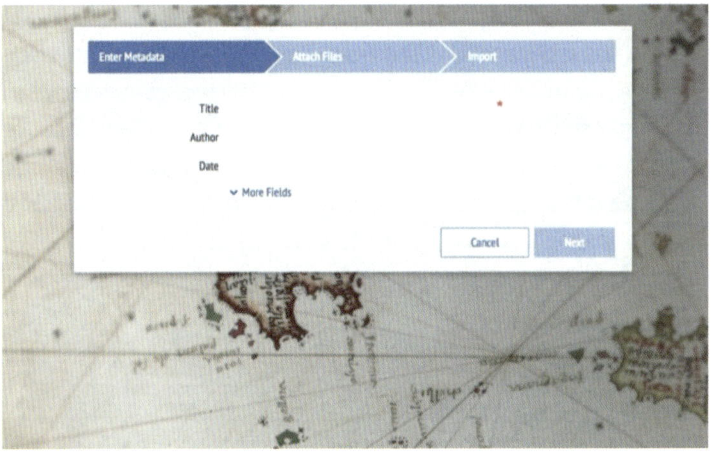

Fig. 6. Screenshot of an upload document page for Recogito.

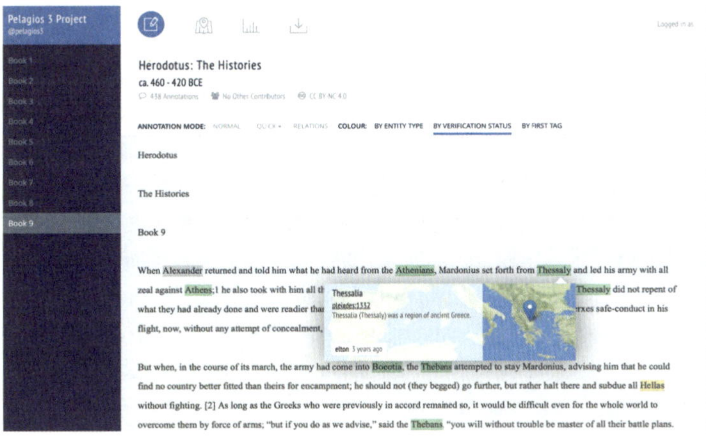

Fig. 7. Screenshot of an annotation page in Recogito, showing a place ("Thessaly") being annotated (with alignment to the Pleiades gazetteer) in Herodotus's *Histories*.

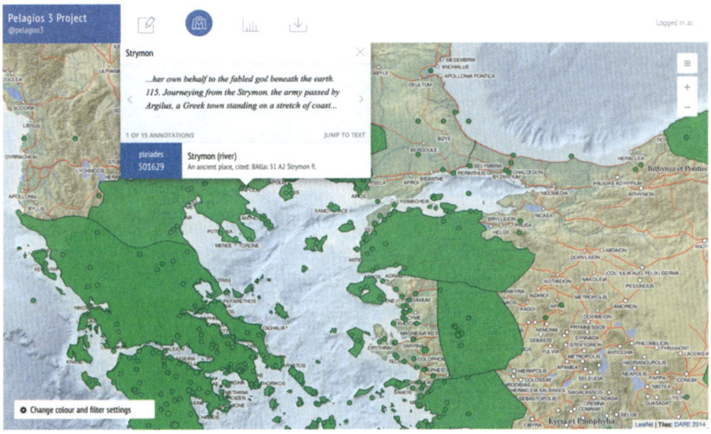

Fig. 8. Screenshot of an output page for Recogito, showing the map of annotations for Herodotus's *Histories*, along with a text snippet for the highlighted place (the river Strymon).

while Recogito was initially designed as a digital tool to extend a technical infrastructure (linked data), first and foremost it is easy to use and puts the onus on the input of the domain expert rather than a data technician. Not only does the Recogito user not need to know anything about linked data; most in fact use Recogito not to produce linked data at all but rather to annotate (and analyze) entities within documents. This aspect of public *use* is critical. Not only have we put the focus on easy use as a way of embedding this kind of practice in current workflows—to make it a "practice of annotation," as it were;[8] we have done so in a way that makes it abundantly clear that every annotation is essentially an assertion or interpretation. Indeed, this is why we believe that semantic annotation should be in the hands of the domain experts. In addition, all annotations are captured in the metadata of Recogito to achieve maximum accountability and transparency.[9]

Producing a technical platform that can be used by anyone, and the results of which are open and transparent to all, needs to be one of the primary concerns of the digital humanities if it is to mature as a discipline that can develop humanistic research and pedagogy in new areas and in new ways. As Lorna Hardwick (in Barker et al. 2012, 191–92) has written, "The potential offered by technology for demolition and constant reconstruction of clusters and systems of knowledge opens sometimes startling associative possibilities. This is both threatening to academic authority and potentially liberating if—and only if the students and other users are equipped to reflect on the new clusters that they create; that is, they do not merely accept these as reflecting a supposed 'natural order' of things but, like Plato in the *Theaetetus*, probe the definition of epistemology as 'true belief together with an account.'" The problematic faith in technology has as much to do with the new tools and methodologies being used, as with the big data that demand "distant reading": as humanists adopt scientific methods to try to make sense of these data, "we also need to be aware of what is at stake in using those digital tools [and] how the tools themselves can distort the research" (Barker et al. 2012, 191–92). As Martin Foys (2013) puts it, "We are still in the age of the digital incunabulum—the time when the new media form has not yet figured out how best to realize its own inherent logic." One of the critical opportunities here is for the humanities to shape the digital research agenda. Arguably the most important advance would be, as Bethany Nowviskie (2013, n.p.) contends, to have better control over the data.

8 With this phrasing, I wish to recall Pierre Bourdieu's (1990) work on *habitus*, both for the sense of agency in the development of infrastructure and for the reproduction of social values in everyday practice. Both points, I believe, are essential for the embedding of Pelagios's linked data infrastructure in the everyday practice of scholarship.

9 While Recogito stores the documents that users upload and the annotations they create, it is not a publication platform. Users are, however, free to make their annotations open, as well as the documents on which they are working, provided that these comply with copyright regulations. Throughout, users remain in control: at any time, they can delete their documents (which will also delete the annotations and everything else connected to the document) or indeed their account from Recogito (which will delete all their document data, metadata, annotations, etc., in compliance with GDPR). See Pelagios, n.d., "Privacy Policy).

If Recogito enables this through putting the power of annotation into the hands of the non-technical user, so that places and spatial entities can be researched in ways far more comprehensive and nuanced than before, how to make use of the resulting linked data annotation network is still only at an early stage of conception and development. The Pelagios team has made an initial foray into this space by developing a prototype search tool called Peripleo (Pelagios. n.d. "Peripleo."; Simon et al. 2016b). As the search engine of linked data maintained by partners of Pelagios, it aims to better enable the discovery and analysis of the connections between scholars, projects, and institutions studying places of the past. Currently most of our partners publish information about ancient places and physical objects, such as archaeological finds, but other kinds of data, including geo-tagged literature or data transcribed from historical maps, as well as other kinds of entities (people and time periods, for example), are also being annotated.

Peripleo enables various search features. First of all is a basic search for places, objects, and other types of data. Using a search box to search by keyword, Peripleo provides suggestions for search terms and specific records matching each query. Search results are coded based on item type, using colors and symbols based on places, objects, time periods, people, and datasets. Second, search results can be narrowed down by using filters. The most prominent part of the filter panel is the time histogram, which shows how search results are distributed over time; but one can also filter by data source, item type, or geographical area. Third, as well as searching for data, summary information about a whole dataset can be displayed. Figure 9 shows an example dataset in Peripleo of an archive, which includes a list of all results contained within its dataset and its geographical footprint on a map.

Peripleo functions rather differently from other map-based search interfaces by virtue of the data in Peripleo being networked. That means items are internally connected through links. One major consequence is how the map works: one dot on the map is not always the same as one search result. Instead, one dot may represent many results connected to the same place, and dot size varies accordingly. Vice versa, one result can appear as many dots. For example, an archaeological artifact might be linked to a findspot as well as to a place of production, while a work of literature might contain references to hundreds of places. Fundamentally, Peripleo enables discovery of data that might be relevant or interesting, based on the idea of place. What it does not do is provide any semantic reasoning for linkages; it is important to note this limitation. As Andrew Prescott (2013) has complained: "There is a tendency to think that if we link a random group of resources together, somehow the magic of linked data will give us instantly new perspectives and new understandings for a particular place or period . . . Sadly, scholarship is much harder than this. Linking of data can be a very useful scholarly technique, but we need to be clear about why we are linking data, what sort of data we are linking, and our aim in doing so." So far as Pelagios is concerned, computers may be linking data, but humans must do the thinking about them.

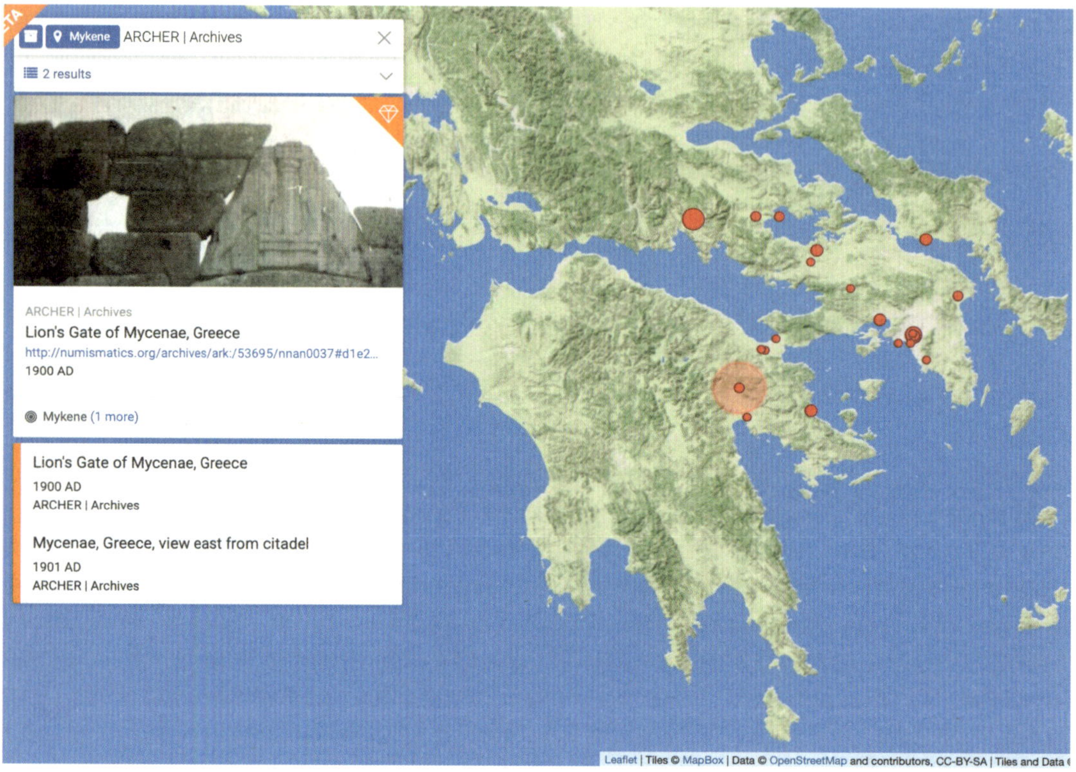

Fig. 9. Screenshot of Peripleo search, showing the spatial coverage of a single collection (the ARCHER archives for coins).

A final point relates to this idea and brings us back, once more, to the community aspect of digital work. We have already seen the social practice of annotation facilitated by Recogito. More broadly, Pelagios's linked data methodology works by linking people as well as data. It primarily seeks to achieve this through our social media account and website, where we are in the process of documenting our resources, especially with regard to how people (from different communities) can use them. Furthermore, a major focus of the current phase of Pelagios has even been on redistributing financial resources (from the grant package that we currently receive from the Andrew W. Mellon Foundation) in the form of Resource Development and Working Groups grants. The former provides grants for the development of resources that will be of benefit for the broader community, while the latter sponsors events for addressing specific issues of a wider concern to the community. One example of a Resource Development Grant has been the redevelopment of the mapping tiles for the Digital Atlas of the Roman Empire (dare.ht.lu.se). Switching from raster-based technology, which sets limits to the area and zoom levels served and consumes a lot of resources, to vector tiles (the information for which is layered, not "baked in"), it has been possible not only to update this map to include different periods but also to make it downloadable for people to use and combine in their own projects.[10] As for working groups, one has been dedicated to multilingualism to translate Pelagios resources into different languages, another to embedding Pelagios tools (especially Recogito) into classroom teaching, while another called "Linked Pasts" has tried to map out the emerging digital ecosystem and what would be needed for this new kind of digital scholarship.[11]

Conclusion

Perhaps due to its fragmentary nature, the study of the ancient world has always been open to a holistic approach that has read archaeology through a historiographical lens and vice versa. At the same time, the growing professionalization of academia, as well as the exponential growth in scholarly output, has resulted in ever more specialized fields of research and ever-sharper disciplinary boundaries being drawn. New digital technologies carry an even greater risk of fragmentation, since the different models and standards used to structure and represent digital data present substantial challenges to their discoverability, accessibility, and (re)usability. And yet, not only can the web be a powerful means of bringing disparate data together for both visualization and analysis, but also scholars of the past now have more vital roles to play in informing the public through this medium.

In this chapter I have explained the process of semantic annotation through two pioneering digital initiatives, the Hestia project and Pelagios, and considered the potential uses of this kind of semantic annotation, including the repurposing of data in other systems (such as GIS or network visualizations) and the linking to other resources. While many challenges remain, I hope to have shown that, contrary to much current thinking, digital tools can in fact enhance, and even demand, the close reading of data in ways that facilitate a more nuanced understanding of place as represented both in our historical sources and through the evidence on the ground. More than this, however, I hope to have highlighted that the social practice of annotation represents one method by which we as experts in our field can communicate information to broader publics with the sufficient robustness, transparency, and accountability necessary for preserving the health of the democratic commons.

10 For a fuller description, see Åhlfeldt 2017a. For a list and description of all Pelagios resource development grants, see Pelagios, n.d., "Resource Development Grants" and "About Working Groups", whose reports are available online via the dedicated Pelagios blog, medium.com/pelagios.

11 Karl Grossner has produced a working "white paper" as a way of making use of the results from the initial work undertaken by the Linked Pasts Working Group (LPWG) of Pelagios. See https://docs.google.com/document/d/1h9pvrsZektiN1OUn03K4smoedjwD4OeW6hfwxijhUW0/edit.

Acknowledgments

I would like to thank Chris Roosevelt and ANAMED for the opportunity to speak before and learn from such an engaged and informed audience. I would also like to thank my colleagues on Hestia (Stefan Bouzarovski, Leif Isaksen, and Chris Pelling) and my investigative team colleagues on Pelagios (Leif again, Rebecca Kahn, Simon Rainer, and Valeria Vitale). The Hestia project was made possible with funding from the UK's Arts and Humanities Research Council (AHRC), in the form of an Early Career Research Grant (2008-2010, ID. No. AH/F019459/1), and with subsequent "follow on" funding (2013-2014), where we shifted from geospatial experimentation to consolidating the technological outcomes and sharing knowledge of those methods and tools with cognate disciplines, interested cultural heritage groups, and the general public. Pelagios has received funding from: JISC (under its Geospatial Engagement and Community Outreach, and Resource and Discovery schemes); the AHRC (Digital Transformations); the Open Knowledge Foundation (Open Humanities Award); and, in three phases of funding, the Andrew W. Mellon Foundation. I benefited also from an Alexander von Humboldt Foundation Research Fellowship for Experienced Researchers (2012-2013) to conduct research with Klaus Geus at the Historische Geographie des antiken Mittelmeerraumes, Freie Universität Berlin and Gerhard Heyer at the Institute of Informatics, University of Leipzig, and an A. S. Onassis Foundation Visiting Fellowship for research at the National Hellenic Research Foundation, Athens (2014-2015).

Works Cited

Åhlfeldt, Johan. 2017a. "Building the Roman Empire Vector Tile Map." Pelagios Commons. November 15, 2017. http://commons.pelagios.org/2017/11/building-the-roman-empire-vector-tile-map/.

———. 2017b. "Digital Atlas of the Roman Empire." https://dare.ht.lu.se/.

Alexandria Archive Institute. n.d. Open Context (website). Accessed May 20, 2019. https://opencontext.org/.

Ancient World Mapping Center. n.d. Antiquity À-la-Carte (website). Accessed May 20, 2019. http://awmc.unc.edu/wordpress/alacarte/.

Bakhtin, Mikhail. 1981. "Forms of Time and of the Chronotope in the Novel: Notes toward a Historical Poetics." In *The Dialogic Imagination: Four Essays by M. M. Bakhtin*, edited by M. Holquist and C. Emerson, 84–258. Austin: University of Texas Press.

Barker, Elton. 2009. *Entering the Agon: Dissent and Authority in Homer, Historiography and Tragedy*. Oxford: Oxford University Press.

Barker, Elton, Christopher Bissell, Lorna Hardwick, Allan Jones, Mia Ridge, and John Wolffe. 2012. "Digital Technologies: Help or Hindrance for the Humanities?" *Arts and Humanities in Higher Education* 11: 185–200.

Barker, Elton, Stefan Bouzarovski, Leif Isaksen, and Chris Pelling, eds. 2016. *New Worlds Out of Old Texts: Revisiting Ancient Space and Place*. Oxford: Oxford University Press.

Barker, Elton, Stefan Bouzarovski, Chris Pelling, and Leif Isaksen. 2010. "Mapping an Ancient Historian in a Digital Age: The Herodotus Encoded Space-Text-Image Archive (HESTIA)." *Leeds International Classical Journal* 9: article no. 1.

Barker, Elton, Stefan Bouzarovski, Chris Pelling, and Leif Isaksen. n.d. Hestia (website). Accessed May 20, 2019. https://hestia.open.ac.uk/.

Barker, Elton, Leif Isaksen, and Jessica Ogden. 2016. "Telling Stories with Maps: Digital Experiments with Herodotean Geography." In *New Worlds from Old Texts: Revisiting Ancient Space and Place*, edited by Elton Barker, Stefan Bouzarovski, Leif Isaksen, and Chris Pelling, 181–224. Oxford: Oxford University Press.

Barker, Elton, and Chris Pelling. 2016. "Space-Travelling in Herodotus 5." In *New Worlds from Old Texts: Revisiting Ancient Space and Place*, edited by Elton Barker, Stefan Bouzarovski, Leif Isaksen, and Chris Pelling, 225–52. Oxford: Oxford University Press.

Barker, Elton, and Melissa Terras. 2016. "Greek Literature, the Digital Humanities, and Shifting Technologies of Reading." Oxford Handbooks Online. https://www.oxfordhandbooks.com/view/10.1093/oxfordhb/9780199935390.001.0001/oxfordhb-9780199935390-e-45.

Beard, Mary. 2017. "Roman Britain in Black and White." *The Times Literary Supplement*, August 3, 2017. https://www.the-tls.co.uk/roman-britain-black-white/.

Bodenhamer, David. 2010. "The Potential of Spatial Humanities." In *The Spatial Humanities: GIS and the Future of Humanities Scholarship*, edited by David Bodenhamer, John Corrigan, and Trevor Harris, 14–30. Bloomington: Indiana University Press.

———. 2015. "Narrating Space and Place." In *Deep Maps and Spatial Narratives*, edited by David Bodenhamer, John Corrigan, and Trevor Harris, 7–27. Bloomington: Indiana University Press.

Bond, Sarah. 2017. "Why We Need to Start Seeing the Classical World in Color." *Hyperallergic*. June 7, 2017. https://web.archive.org/save/https://hyperallergic.com/383776/why-we-need-to-start-seeing-the-classical-world-in-color/.

Bourdieu, Pierre. 1990. *The Logic of Practice*. Cambridge: Polity Press.

Bouzarovski, Stefan, and Elton Barker. 2016. "Between East and West: Movements and Transformations in Herodotean Topology." In *New Worlds from Old Texts: Revisiting Ancient Space and Place*, edited by Elton Barker, Stefan Bouzarovski, Leif Isaksen, and Chris Pelling, 155–79. Oxford: Oxford University Press.

Day of Archaeology. 2017. "Find Out What Archaeologists Really Do." http://dayofarchaeology.com.

de Certeau, Michel. 1984. *The Practice of Everyday Life*. Berkeley: University of California Press.

Dear, Michael, Jim Ketchum, Sarah Luria, and Douglas Richardson, eds. 2011. *GeoHumanities: Art, History, Text at the Edge of Place*. New York: Routledge.

Detienne, Marcel. 1996. *The Masters of the Truth in Archaic Greece*. New York: Zone Books.

Dewald, Carolyn. 1987. "Narrative Surface and Authorial Voice in Herodotus' Histories." *Arethusa* 20: 147–70.

Dörk, Marian, Rob Comber, and Martyn Dade-Robertson. 2014. "Monadic Exploration: Seeing the Whole through Its Parts." In *Proceedings of the SIGCHI Conference on Human Factors in Computing Systems, Toronto, Canada*. New York: ACM. http://dx.doi.org/10.1145/2556288.2557083.

FORCE11. 2017. "The FAIR Data Principles." https://www.force11.org/group/fairgroup/fairprinciples.

Foys, Martin. 2013. "How I Learned to Stop Worrying and Love Big Data." *Burnable Books* (blog). https://web.archive.org/web/20160205024906/http://burnablebooks.com/foysonbigdata/ (discontinued).

Free Press. n.d. "Save the Internet: Net Neutrality." Accessed August 19, 2019. https://www.freepress.net/issues/free-open-internet/net-neutrality.

Futo-Kennedy, Rebecca. 2017. "Classically White Supremacy: The American Dream of a White City." *Classics at the Intersection* (blog). December 8, 2017. https://web.archive.org/web/20190328075111/https://rfkclassics.blogspot.com/2017/12/classically-white-supremacy-american.html.

Goldhill, Simon. 1986. *Reading Greek Tragedy*. Cambridge: Cambridge University Press.

Graham, Shawn, Neha Gupta, Jolene Smith, Andreas Angourakis, Michael Carter, and Beth Compton. n.d. "The Open Digital Archaeology Textbook." Accessed May 20, 2019. https://o-date.github.io/.

Gregory, Ian, and David Cooper. 2010. "GIS, Texts and Images: New Approaches." *Poetess Archive Journal* 2, no. 1.

Gregory, Ian, and Richard Healey. 2007. "Historical GIS: Structuring, Mapping and Analyzing Geographies of the Past." *Progress in Human Geography* 31, 638–53.

Gregory-Guider, C. 2004. "'Deep Maps': William Least Heat-Moon's Psychogeographic Cartographies." *eSharp-electronic Journal* 4, 1–17.

Gurevich, Aron. 1985. *Categories of Medieval Culture*. London: Routledge.

Harley, J. B. 1989. "Deconstructing the Map." *Cartographica* 26, 1–20.

Harris, Trevor, Susan Bergeron, and L. Jesse Rouse. 2011. "Humanities GIS: Place, Spatial Storytelling, and Immersive Visualization in the Humanities." In *GeoHumanities: Art, History, Text at the Edge of Place*, edited by Michael Dear, Jim Ketchum, Sarah Luria, and Douglas Richardson, 226–40. New York: Routledge.

Harvey, David. 1996. *Justice, Nature and the Geography of Difference*. Oxford: Wiley-Blackwell.

Janni, Pietro. 1984. *La Mappa e il Periplo: Cartogra a antica e spazio odologico*. Rome: Giorgio Bretschneider.

Kerferd, G. B. 1981. *The Sophistic Movement*. Cambridge: Cambridge University Press.

Kurke, Leslie. 1999. *Coins, Bodies, Games, and Gold: The Politics of Meaning in Archaic Greece*. Princeton: Princeton University Press.

Lefebvre, Henri. 1991. *The Production of Space*. Oxford: Wiley-Blackwell.

Levine, Caroline. 2015. *Forms: Whole, Rhythm, Hierarchy, Network*. Princeton: Princeton University Press.

Malkin, Irad. 2011. *A Small Greek World: Networks in the Ancient Mediterranean*. Oxford: Oxford University Press.

Massey, Doreen. 2005. *For Space*. London: Sage Publications.

Moretti, Franco. 1998. *Atlas of the European Novel 1800–1900*. London: Verso.

———. 2011. "Network Theory, Plot Analysis." *Stanford Literary Lab Pamphlet* (SLLP) 2, 2–31.

Morley, Neville. 2017. "Diversitas et Multiculturalismus." *Sphinx* (blog). August 2, 2017. https://web.archive.org/save/https://thesphinxblog.com/2017/08/02/diversitas-et-multiculturalismus/.

Nowviskie, Bethany. 2013. "Resistance in the Materials." January 4, 2013. https://web.archive.org/web/20190328075221/http://nowviskie.org/2013/resistance-in-the-materials/.

Pelagios. n.d. “About Working Groups.” Accessed May 20, 2019. https://web.archive.org/web/20190329131234/http://commons.pelagios.org/working-groups/.

———. n.d. “Pelagios Commons.” Accessed May 20, 2019. http://commons.pelagios.org/.

———. n.d. “Peripleo.” Accessed May 20, 2019. https://peripleo.pelagios.org/.

———. n.d. “Privacy Policy.” Recogito Help. Accessed May 20, 2019. https://recogito.pelagios.org/help/privacy.

———. n.d. “Recogito.” Accessed May 20, 2019. https://recogito.pelagios.org/.

———. n.d. “Resource Development Grants.” Accessed May 20, 2019. https://web.archive.org/web/20190329131218/http://commons.pelagios.org/resource-development-grants.

“Pelagios.” n.d. GitHub. Accessed May 20, 2019. https://github.com/pelagios.

Pelagios (blog). n.d. Accessed August 19, 2019. https://medium.com/pelagios.

Pelling, C. B. R. 1997. “East Is East and West Is West—or Are They?: National Stereotyping in Herodotus.” *Histos* 1, 51–66.

Perseus Digital Library. n.d. Accessed May 20, 2019. http://www.perseus.tufts.edu/hopper/.

Pred, Allan. 1983. “Structuration and Place: On the Becoming of Sense of Place and Structure of Feeling.” *Journal for the Theory of Social Behaviour* 13 (1), 45–68.

Prescott, Andrew. 2013. The Function, Structure and Future of Catalogues.” *Digital Riffs* (blog). January 11, 2013. https://web.archive.org/save/http://digitalriffs.blogspot.com/2013/01/the-function-structure-and-future-of.html.

Purves, Alex C. 2010. *Space and Time in Ancient Greek Narrative*. Cambridge: Cambridge University Press.

Ridge, Mia, Don Lafreniere, and Scott Nesbit. 2013. “Creating Deep Maps and Spatial Narratives through Design.” *International Journal of Humanities and Arts Computing* 7, 176–89.

Said, Edward. 1978. *Orientalism*. New York: Pantheon Books.

Scheidel, Walter, and Elijah Meeks. n.d. “ORBIS: The Stanford Geospatial Network Model of the Roman World.” Accessed May 20, 2019. http://orbis.stanford.edu/.

Simon, Ranier, Elton Barker, Leif Isaksen, and Pau de Soto Cañamares. 2017. “Linked Data Annotation Without the Pointy Brackets: Introducing Recogito 2.” *Journal of Map & Geography Libraries* 13 (1): 111–32. https://doi.org/10.1080/15420353.2017.1307303.

Simon, Rainer, Leif Isaksen, Elton T. E. Barker, and Pau de Soto Cañamares. 2016. “The Pleiades Gazetteer and the Pelagios Project.” In *Placing Names: Enriching and Integrating Gazetteers*, edited by Merrick Lex Berman, Ruth Mostern, and Humphrey Southall, 97–109. Bloomington: Indiana University Press.

Tuan, Yi-Fu. 1977. *Space and Place: The Perspective of Experience*. Minneapolis: The University of Minnesota Press.

Turnbull, David. 2007. "Maps, Narratives and Trails: Performativity, Hodology and Distributed Knowledges in Complex Adaptive Systems—an Approach to Emergent Mapping." *Geographical Research* 45, 140–49.

Vernant, Jean-Pierre. (1962) 1982. *The Origins of Greek Thought*. Ithaca: Cornell University Press.

Wood, Denis. 2010. *Rethinking the Power of Maps*. New York: Guilford Press.

Zuckerberg, Donna. 2017. "Learn Some F*cking History." *Eidolon* (blog). October 5, 2017. https://eidolon.pub/learn-some-f-cking-history-94f9a02041d3

II

Mapping Archaeology

Intra-Site GIS at Çatalhöyük (2009–2017): From an Integrative and Reflexive Tool to the Living Archive

Camilla Mazzucato

Claudia Engel

Dominik Lukas

> *At Çatal Höyük there is no short cut to obtaining quick results and the only course for action is that of the gradual stripping off layer after layer.*
> — James Mellart, "Excavations at Çatal Hüyük, 1963, Third Preliminary Report"

Introduction

At the end of August 2017, after twenty-five years of intense research, the Çatalhöyük Research Project (ÇRP) concluded its longtime commitment to the excavation, conservation, and study of the Neolithic site of Çatalhöyük. The aim of this chapter is to sketch the journey of Çatalhöyük's intra-site GIS from its beginnings in 2009 to the end of ÇRP's excavation work. We intend to highlight its role within the different stages of the ÇRP excavation process, focusing especially on its function as a tool for data integration that can enhance reflexivity and foster multidisciplinary collaboration between researchers (Berggren et al. 2015). We then intend to focus on the final incarnation of the ÇRP GIS project as part of the Living Archive of Çatalhöyük (LAÇ) web application. In particular, we will illustrate how Çatalhöyük's diverse digital data have been harmonized and made accessible through the LAÇ. We will, moreover, highlight plans for future development of the LAÇ and delineate a series of suggestions and best practice principles when developing such systems. In this way we hope to contribute to current critical discussions on how to archive effectively, how to make digital archaeological data accessible and reusable, and how to harmonize diverse and

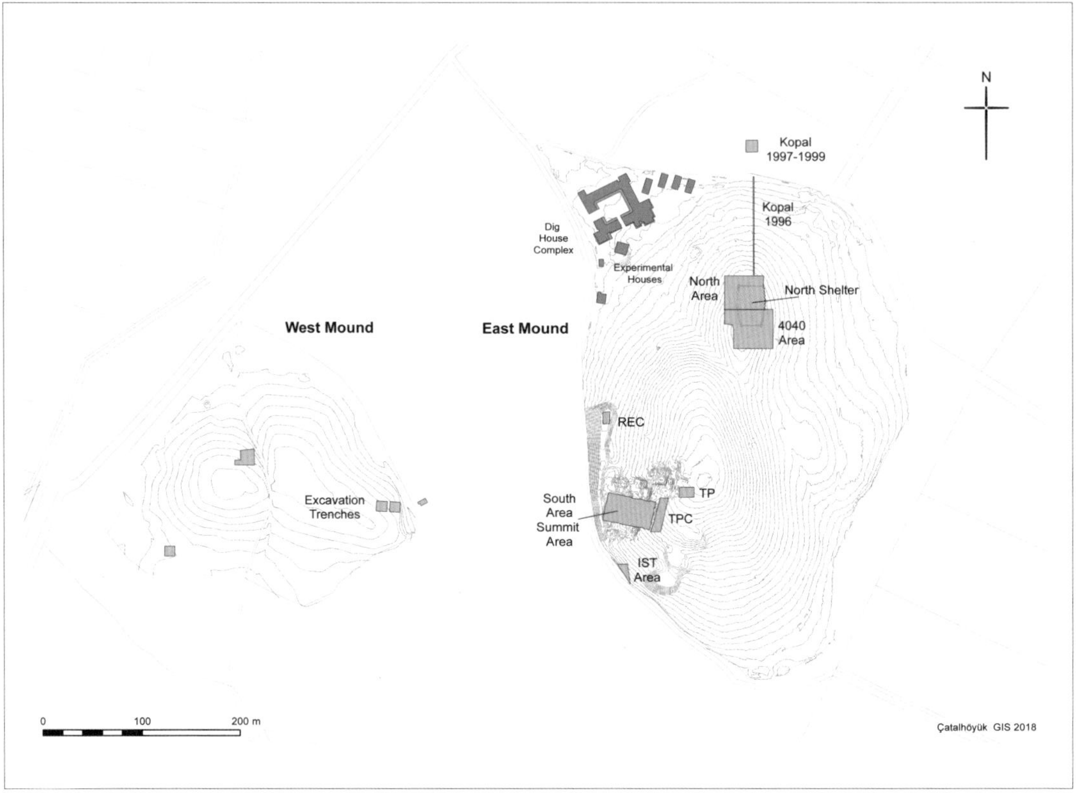

Fig. 1. Çatalhöyük East and West Mounds with Çatalhöyük Research Project excavation areas (Courtesy of ÇRP).

fragmented archaeological datasets productively (see Atici et al. 2013; Barker et al. 2010; Engel and Grossner 2015; Kansa 2012; Kintigh et al. 2015; Isaksen et al. 2010).

Reflexivity: the "*Fil Rouge*" and Data Complexity

Çatalhöyük is an extremely complex, long-lived, multilayered archaeological site. Declared a UNESCO World Heritage Site in 2012, the large settlement of Çatalhöyük spread during its lifetime over two small artificial mounds located in the southwest area of the Konya Plain, central Anatolia (Fig. 1). The East Mound, containing the core Neolithic occupation of the site, reached a maximum extent of approximately 13.5 hectares and was inhabited with a considerable degree of continuity for more than a millennium (7100–6000 BCE) (Bayliss et al. 2015; Hodder 2014). The size of the population is debated, but estimates point to a population of between three thousand five hundred and eight thousand people (Cessford 2005), making Çatalhöyük's East Mound one of the largest Neolithic settlement agglomerations in the Near East.

To date, the deep stratigraphic sequence of the Çatalhöyük East Mound has been extensively excavated by two projects. The first was led by James Mellaart, who worked mainly in the southwestern part of the East Mound at Çatalhöyük between 1961 and 1965. Over four seasons of work, Mellaart and his team managed to unearth more than 160 buildings (Mellaart 1962, 1963, 1964, 1966, 1967), exposing a unique, undisturbed, and continuous occupation of overlapping mudbrick buildings and a repertoire of material culture and symbolic representations of a richness unmatched for this period. Although the project did not last very long, these excavations had the effect of dramatically changing the way the prehistoric archaeology of central Anatolia was perceived, bringing the site to the forefront as an important node in the spread of Neolithic lifeways to Anatolia.

After decades of abandonment, a new excavation project directed by Ian Hodder was established in 1993: the ÇRP. The new project was created with the intention of working twenty-five years, with a varied and wide set of new research objectives open to a new array of stakeholders and involving international teams and a range of specialists (Hodder 1996; Farid and Hodder 2014; Hodder, Cessford, and Farid 2007). The ÇRP expanded greatly the excavation horizon and, while the work in the southwestern part of the East Mound was carried on (the South Area), a number of new excavation areas were opened to answer specific research questions (Fig. 2).

During the 2014 field season, Ian Hodder (2015) counted 160 team members divided between field archaeologists and laboratory specialists, encompassing 36 different specialties. The diversity of methodologies put in place by the ÇRP over the years—together with the amount and variety of scholars involved in different stages of research as well as the theoretical, temporal, and financial engagements with the investigation and preservation of the site—led the ÇRP to be considered one of the most "comprehensive and complex archaeological projects in contemporary archaeology" (Marciniak 2015, 3; Farid 2015). The size and complexity of the ÇRP's research structure is reflected in the amount and diversity of data the project generated during its activity. During its twenty-five-year run, the ÇRP produced a "data deluge" (Faniel and Zimmerman 2011; Bevan 2015) that challenges storage practices, as well as analytical and visualization processes. As an active project running over the last three decades, the ÇRP served as a laboratory for many developments in digital archaeology—those producing data types still in use today, but also those produced with software long since deprecated. Data types range from a variety of text documents (DOC, RTF, PDF, ODF, etc.), spreadsheets (XLS, DBF, etc.), raster (BMP, TIF, JPG, GIF, etc.) and vector images (DXF, DWG, AI, EPS, etc.), and moving images (AVI, FLV, MOV, MP4, etc.) to more exotic files no longer in use today. While this multiplicity of media produced genuinely "thick description" of the site, it frequently resulted in a dispersed and disconnected archaeological documentation process where different data types often belonged to different parts of the physical or digital archive and were frequently difficult to reconnect (Engel and Grossner 2015; Marciniak 2015; Mickel 2015).

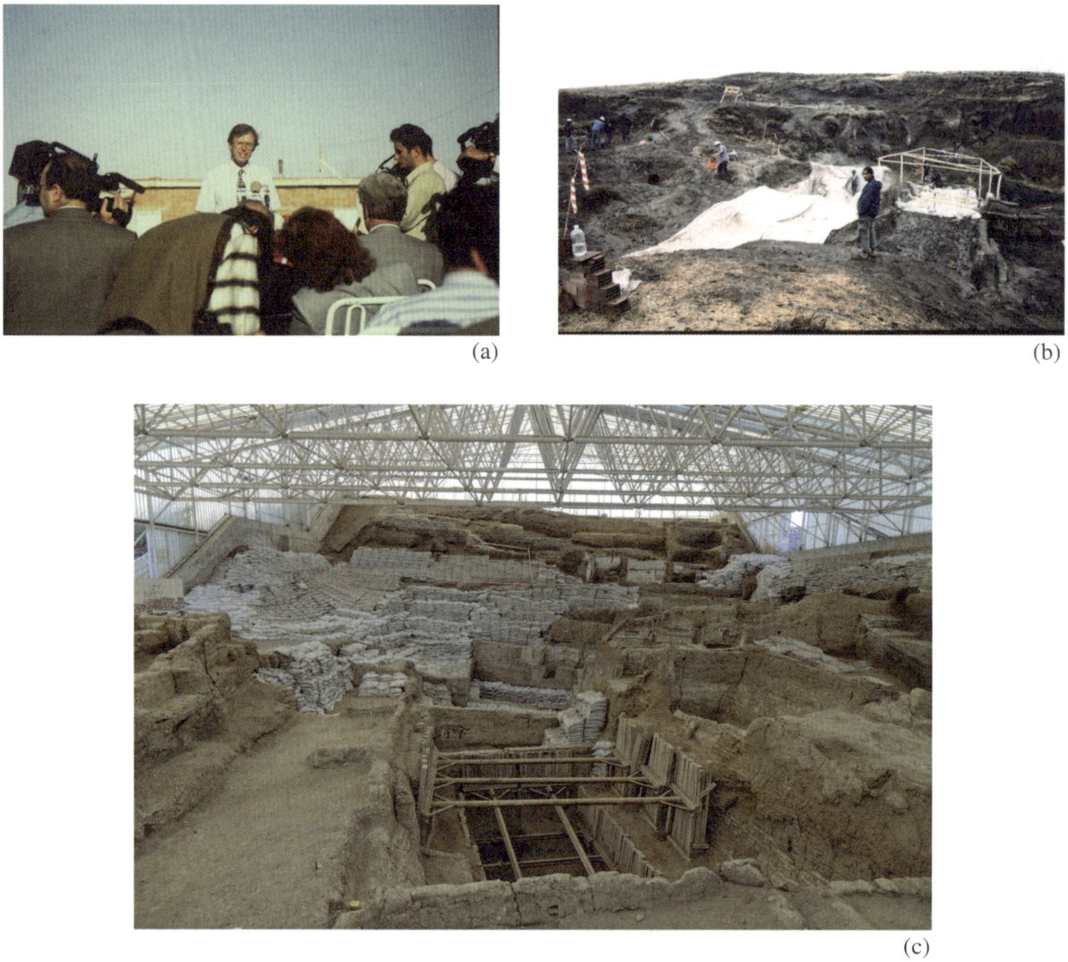

Fig. 2. (a) Reopening the site. Ian Hodder at a press conference, 1996 (Courtesy of Burcu Tung). (b) Shahina Farid in the South Area, 1997 (Courtesy of Burcu Tung). (c) The South Area at the end of the 2017 season (Courtesy of Scott Haddow).

This complexity is further increased by the presence, as a vital part of the ÇRP's digital archive, of data produced through reflexive engagements on-site that informed all phases of the ÇRP's lifespan (Lukas, Engel, and Mazzucato 2018). The ÇRP was conceived with the main methodological objective of putting into practice a reflexive method of archaeological research rooted in a post-processual theoretical approach to archaeological knowledge production (Berggren and Hodder 2003; Berggren and Nilsson 2014; Hodder 1997, 2000b, 2003). The strong commitment to critical reflexivity, which embodies the raison d'être of the project, persisted through the closure of the project as the most significant and distinctive feature of the ÇRP and endured despite frequent changes in team members, technological tools, research agendas, and interpretations (Berggren and Nilsson 2014; Berggren et al. 2015).

Central to the reflexive approach is the idea of the substantial faultiness of compartmentalized archaeological practices and the strong commitment to breaking down barriers, first and foremost, between field recording and later interpretation (Berggren and Nilsson 2014). Archaeological research is envisioned as open and transparent, and archaeological knowledge is always regarded as situated and contingent; it should freely flow within a network of diverse voices all equally entitled to be part of the discussion (Hodder 1997, 2000, 2003, 58). This multivocal nature of knowledge production should be achieved from the very beginning of excavation process, that is, at the "trowel's edge" (Hodder 2003, 58). It is at the "trowel's edge" that archaeologists start interpreting their deposits, and this happens in light of their own experience and contingent positionalities. Indeed, the reflexive framework entails the recognition of one's positionality and the way in which this affects knowledge creation throughout the entire course of research (Hodder 2003). Team members are encouraged to be transparent about their interpretations, and a key part of the research process involves finding time for and incorporating tools for discussions about the interpretation process (Berggren and Hodder 2003; Hodder 2003). Since the 1990s the ÇRP has put a lot of effort in developing an efficient, reflexive workflow at the site (for a detailed description, see Hodder 2000; Berggren and Nilsson 2014). Team members, from both the excavation and laboratory teams, were encouraged to provide detailed accounts of their interpretation process and context, of possible uncertainties, alternative interpretations, and competing ideas. Specific locations of the recording system were dedicated spaces for reflexive accounts: some were free-text fields in the database, like unit description and interpretation fields, or the diaries (Mickel 2015). Videos and priority tours and, more recently, digital sketches drawn on tablet computers were also part of the reflexive apparatus (Berggren et al. 2015; Berggren and Nilsson 2014).

Digital technologies had an important role in the way the ÇRP functioned; since the reopening of the site in 1993, the most up-to-date digital technologies formed an integral part of the site's recording protocols (Berggren et al. 2015; Farid 2015). Such technologies were changed, were updated, and were supplemented by new additions to existing archaeological practices over the years, as a result of continual negotiations between team members, research objectives, and interpretative approaches. These negotiations were frequently complicated and non-linear, with digital methods transforming existing archaeological practices and being transformed by them. Part of this dialectical process left traces in the site's digital archive, which itself is a source of reflexive data. Additionally, the implications of new digital tools and the way these fit within archaeological practice and influence processes of knowledge creation were always part of the reflexive ethos at the site (see Berggren and Hodder 2003). A number of recent articles have highlighted a series of aspects of this reflexive engagement with digital recording and analytical methods (Berggren et al. 2015; Lukas, Engel, and Mazzucato 2018; Taylor et al. 2018).

While the ÇRP had the inclination and, perhaps more importantly, the resources to experiment with digital technologies and new recording methods, the complex and long-lived nature of the project meant that earlier decisions regarding technology and archaeological practices frequently constrained subsequent choices. Over the years, the intra-site GIS was part of this thick network of technology and practice. When the intra-site GIS was created, ESRI's ArcGIS software suite was chosen as the most convenient option given the existing technological infrastructure. ArcGIS offered necessary capabilities for linking to the project's already well-developed and centralized Microsoft Access database. Additionally, the robust single-context excavation methodology employed by the ÇRP, based on MoLAS standards, afforded the implementation of a unit-based GIS system where each unit was stored as a vector polygon, linked to a selection of textual information in the excavation database.

We will now trace the development of the intra-site GIS through three macro-phases of the ÇRP: (1) the implementation phase; (2) the subsequent shift from a tool for off-site/post-excavation processing to one for on-site primary data recording; and (3) the final stage as part of the LAÇ.

GIS as a Reflexive Tool

The ÇRP's GIS project was established in 2009, and it was built on an ecosystem of existing digital data sets, tools, and well-established relations from the beginning. It relied mainly on the project's centralized database, which provided the GIS with tabular information via a dynamic link, and on the project's existing and advanced digital cartographic practices, which had produced digital representations of all excavated units using AutoCAD since the early 2000s. Within the "archaeological assemblage" the intra-site GIS quickly become a central tool that partially replaced older tools (e.g., AutoCAD), and turned into the site's primary spatial data storage, as well as a crucial instrument for primary and secondary data capture.

In the existing ecosystem, the new GIS project was conceived in order to play an important integrative role. Its adoption aimed to articulate segments of the ÇRP's archaeological research that still remained partially fragmented in a better way and to advance the process of "developing of a common language," a process that accelerated with the 2004 implementation of the project's centralized database infrastructure. The centralized database system was made available to all team members during both the excavation and off-site periods (Engel and Grossner 2014; Ridge and May 2004; Ridge, May, and Mackie 2005).

The ÇRP intra-site GIS was conceived as a "flexible link between work conducted in the field and work carried out by specialists, between the graphic and spatial information produced by excavators and the textual information stored in the central database" (Berggren et al. 2015; Taylor et al. 2018, 1/7). Additionally, the integrative capabilities of GIS were valued for the software's ability to store, display, and analyze data generated by different specialists within a single environment. This capability raised the possibility of a more fluid and efficient multidisciplinary research program and a more interactive engagement with data (Berggren et al. 2015; Taylor et al. 2015). The intra-site GIS was seen as a tool capable of facilitating the work of interlacing, braiding, and assembling different data, theories, and people that constitute the way knowledge of the past was constructed at the site (Hodder 2015; Marciniak 2015, 3).

The ÇRP intra-site GIS was designed with the intention of promoting a series of digital practices that would be fully integrated within the project's reflexive theoretical framework and that would enhance the reflexive ethos on-site. From the beginning it developed as a collaborative and bottom-up effort, with the contribution of archaeologists and specialists who were variously involved in different stages of the construction of the GIS geodatabase that formed the core architecture and principal data repository of the system (Taylor et al. 2018, 1/9). The outcome of this collaborative process was a tool tailored for the site, a highly flexible and contextual system that changed and grew in accordance with the needs of archaeological practice at the site (Taylor et al. 2018, 1/7). While the core of the GIS geodatabase—a synthesis of the excavation database, wherein each excavated and recorded unit was drawn as a vector polygon—was never modified, other parts were expanded or pared down as the practical and analytical needs of excavators and specialists changed over time. Consequently, the ÇRP GIS was always a work in progress: never complete or static but always in the process of being updated (Edgeworth 2014).

GIS practices at the site unfolded in a frequently untidy manner, in fact, creating numerous moments of tension. Tensions resulted in part from the way the system was constructed, requiring adjustments over time, similar to the changes required of the centralized database through the years (Lukas, Engel, and Mazzucato 2018). The integration within the geodatabase of 3-D models created through a structure from motion (SfM) technique produced one of these moments, because of problems inherent in incorporating multifeature/unit models within a unit-based GIS structure (Berggren et al. 2015; Forte et al. 2015; Taylor et al. 2018; for 3-D model integration within GIS, see also Dell'Unto et al. 2015; Landeschi et al. 2016).

An issue that quickly emerged was that the ÇRP's commitment to multivocality conflicted with the way the GIS and the centralized relational database functioned because they required high levels of standardization to work smoothly. The development of a "common language" for knowledge sharing between team members, however, always remained part of ÇRP's goals. The free flow of knowledge required a shared framework and a standardized way of recording the archaeology (Hodder 2000; Lucas 2012). In this regard the single-context, on-site recording system offered a solid basis for interpretation of data. This system endured as the foundation of the excavation work of the ÇRP, despite some adjustments to adapt it to the complexity of the Neolithic archaeology (e.g., the creation of feature entities composed of grouped units) (Farid 2015; Farid and Hodder 2014).

The question of whether GIS can (or should) complement and enhance a reflexive agenda has always been controversial. GIS research has traditionally been seen as adhering to a positivist, modernist spatial-analytical approach. The conflation of GIS research with positivism and modernity led to an almost complete rejection of the software by researchers practicing phenomenological approaches. A non-representational-thinking perspective on the role of GIS within archaeological research seems a fruitful lens through which to view its utility as part of a reflexive ethos (see Hacıgüzeller 2012, 2017 for an in-depth critique).

At Çatalhöyük, digital GIS practices and representations fully participate in the production of archaeological knowledge and reside at the center of the ÇRP recording system, fully integrated within the reflexive assemblage. Analyses conducted within the ÇRP GIS, and representations generated from it, are part of a fully integrated ecology of mixed media: a plurality of tools and representations that allow for a "multivocality" of practice.

Going Paperless: GIS at the Trowel's Edge

The shift to fully digital recording practices at Çatalhöyük started quite late compared to other archaeological projects, many of which had already adopted an advanced practice of digital recording on-site (e.g., Hartzler 2009; Jackson, Motz, and Brown 2016; Olson et al. 2013; Opitz 2014; Roosevelt et al. 2015). The slow adoption of paperless recording resulted mainly from the size and complexities of the ÇRP system. The first experiments in moving to a fully paperless recording system began in 2014. Microsoft tablet computers were chosen for recording primary data in the field. These tablets were used as flexible and portable "hubs," as a means of placing the integrative capabilities of the intra-site GIS into the excavation trench (Fig. 3). A series of dedicated vector GIS files (shapefiles) were created, uploaded to the tablets, and used as repositories for spatial data recorded on-site (for a critical assessment of the move to digital recording of primary data, see Taylor et al. 2018).

The goal of adopting a fully digital recording practice was achieved in 2015: the intra-site network was extended to the North and South excavation areas, and the centralized database was brought to the trench via the tablets. Archaeologists could enter textual information in the centralized database at the trowel's edge. This move allowed the field team to reduce the distance between the archaeology and the digital recording (Taylor et al. 2018).

The GIS changed from a cartographic and analytical tool used mainly in the lab and outside the excavation area for the processing of excavation data into a tool for

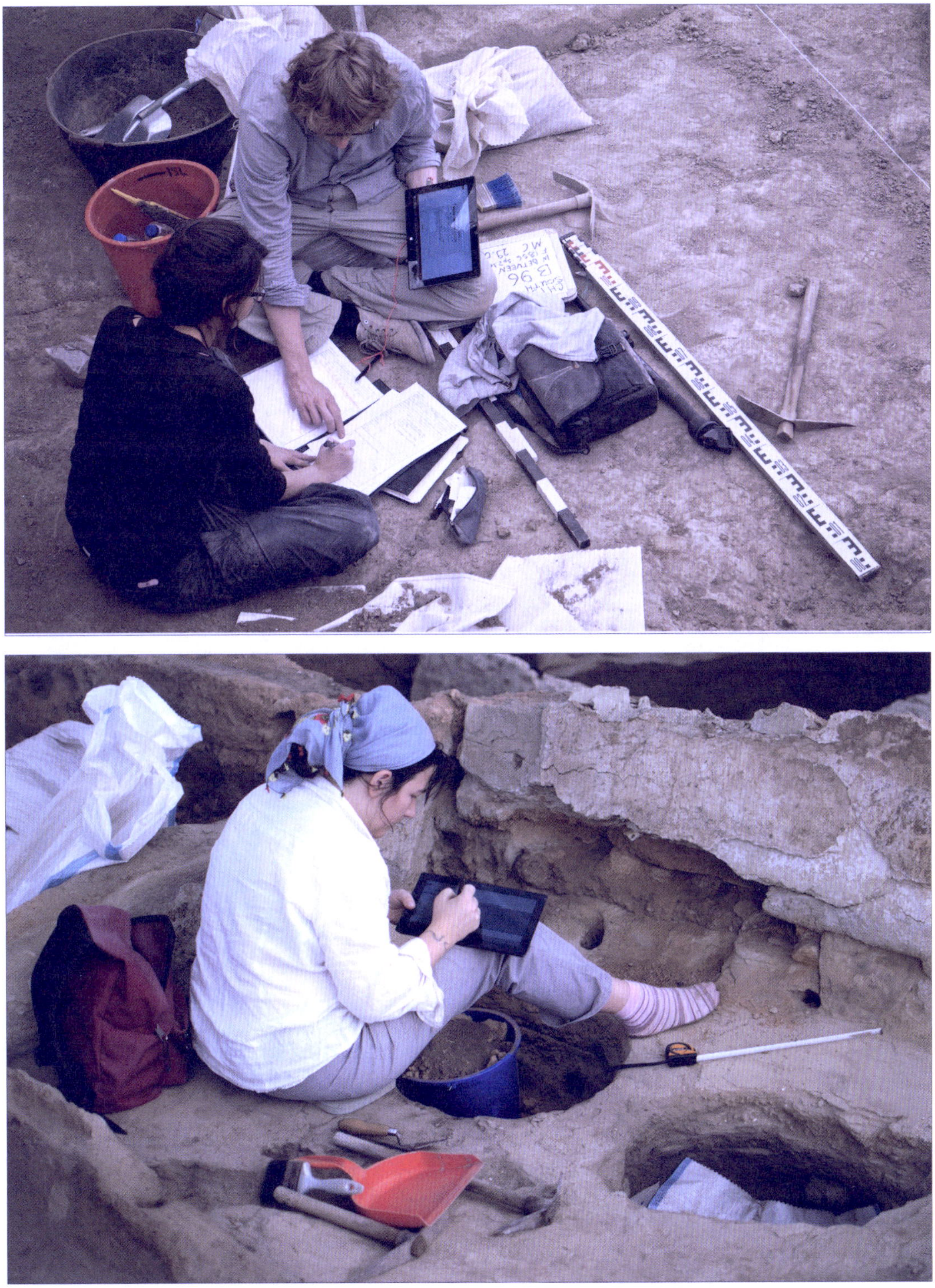

Fig. 3. Using tablet computers for primary data recording on site (Courtesy of ÇRP).

primary data recording in the excavation trench. Before this shift, much of the GIS work was accomplished only after the excavation season, frequently resulting in "out-of-date" digital records. The move to digital recording was, in essence, a change in the nature of GIS work on excavation data: from a process of transcoding (the translation of media into different media; Manovich 2001) paper field drawings to the production of a fully "born digital" archaeological record. This shift allowed the digital record to remain current—for the most part—with excavation activity.

As a result of these changes, the master GIS geodatabase had to be adjusted; new parts were added to host the shapefiles coming from field tablets. A new "tablet" feature dataset was organized in a series of feature classes that mirrored on-site recording practices and their associated shapefiles used for data recording. Information previously contained in paper field drawings was atomized and divided into a series of shapefiles in an effort to preserve the information richness of paper field drawings (Fig. 4; Taylor et al. 2018). Additionally, a set of new elements and geometries were included in the geodatabase; a feature class was created to host the details added to paper field drawings to characterize units, for instance. These details had usually not been included in the geodatabase previously, but it was deemed important to preserve such information with the migration to a fully "born digital" recording system (for a critical review of the move from paper to digital field drawings, see Hacıgüzeller 2017 and Morgan and Wright 2018).

The paper-based recording routine proved difficult to replace, and we soon came to realize that ArcGIS was ill-suited to drawing archaeological sections on-

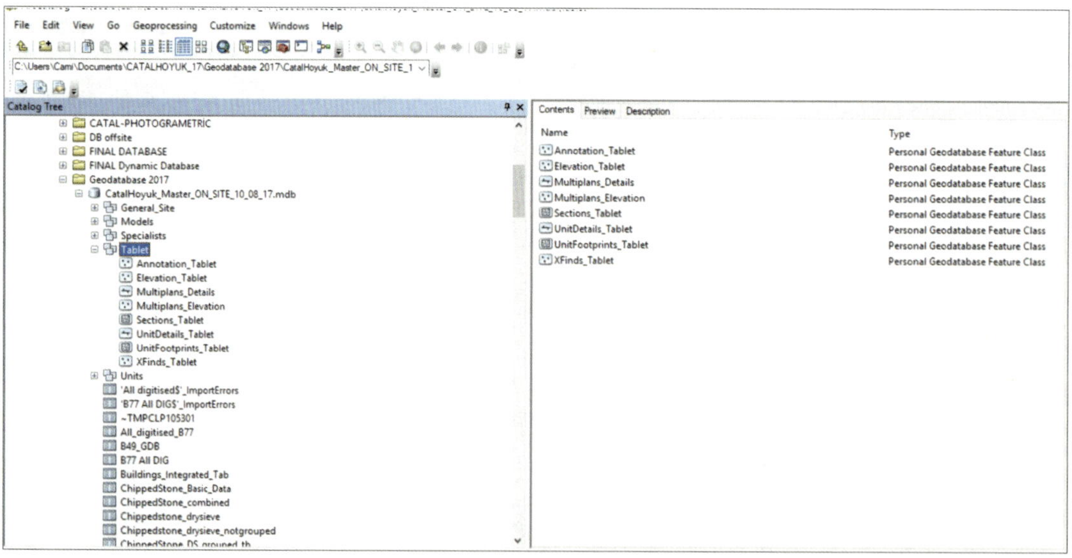

Fig. 4. Geodatabase tablet feature dataset structure. The feature classes are as follows: UnitFootprints_Tablet: storing all the units as polygons; Elevation_Tablet: storing elevations for each unit as points; Annotation_Tablet: storing selected notes of interest as points; XFinds_Tablet: storing X finds coordinates as points; and UnitDetails_Tablet: storing details and characterizations for each unit as lines.

site. Nonetheless, tablet computers enhanced the reflexive capabilities of the GIS by shortening the distance between archaeological recording and the digital archive (see Taylor et al. 2018).

The Living Archive of Çatalhöyük (LAÇ)

With the end of excavations in 2017, the problem of how to store and make easily available and reusable the enormous body of data produced by the project over the years became increasingly urgent. Furthermore, the ÇRP's recording system had grown significantly over time and the complex stratification of datasets and technological infrastructures led to a very fragmented process of data creation and retrieval. As Engel and Grossner (2014, 17) noticed, "The (current) system [was] tuned primarily to putting data in rather than getting it out." Retrieving data from the ÇRP system requires a deep knowledge of its structure and history and, frequently, advanced technical skills. Given the size and complex structure of the project, the harmonization of diverse data had always been an issue.

A pilot project of the LAÇ was inaugurated in 2014 (Fig. 5) with a series of specific objectives: ensuring permanent accessibility to ÇRP's datasets after the end of the project in 2017; supporting a reflexive approach to the study of Çatalhöyük's archaeological record by integrating annotations bolstered by particular queries; making the vast knowledge stored in the digital archive transparent and accessible by more efficiently re-connecting the reflexive layer of information to the rest of the datasets; and allowing for semantic queries of datasets by exposing the data as RDF triples (Engel and Grossner 2015).

The LAÇ developed within an ongoing conversation regarding the interlinked issues of long-term curation, preservation, and reutilization of digital archaeological datasets, as well as the strategies required to improve interoperability between

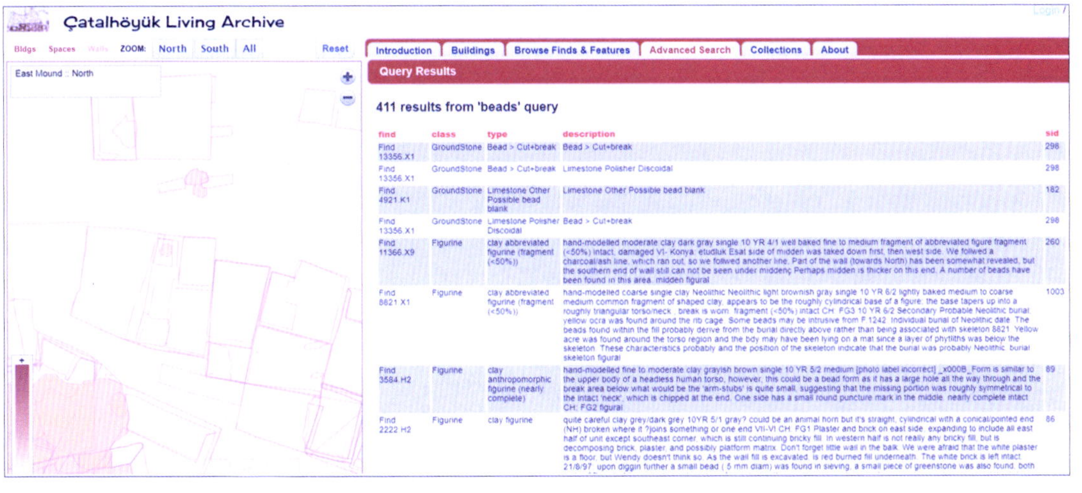

Fig. 5. The Living Archive Çatalhöyük (LAÇ) web application (screenshot).

archaeological recording systems, openness, and sharing of data (see Atici et al. 2013; Faniel and Zimmerman 2011; Faniel et al. 2013; Isaksen et al. 2010, 2014; Kansa 2012).

Currently, archaeological projects that want to put in place a plan for long-term archiving of digital data have the option of submitting them to agencies like the Archaeology Data Service (ADS; University of York) or the Digital Archaeological Record (tDAR; University of Arizona) that offer static types of data archiving. While these agencies ensure preservation, future readability, and retrieval of data, they also provide ways to integrate data archives with ongoing or future research. Furthermore, these archives require that the internal relationships of relational data be broken and atomized (i.e., relational database tables must be archived as individual spreadsheets), thus making complex data queries largely impossible (Lukas, Engel, and Mazzucato 2018).

The LAÇ, in contrast, was conceived as a dynamic and open tool, available to researchers and the public, that aims to keep ÇRP data highly accessible and, most importantly, available for the "re-interpretation of objects and the re-assembling of their contexts" (Engel and Grossner 2015, 18). Moreover, a fundamental aim of the LAÇ is to make the reflexiveness and multivocality of the research process transparent through a close integration of the various datasets, their provenance, relationships, and contexts. To this end the use of ontologies and semantic web technologies open up opportunities for the design of a flexible framework for data integration, storage, and reutilization (Isaksen 2011; Lukas, Engel, and Mazzucato 2018).

Through the implementation of semantic web ontologies, we developed a formal specification of shared conceptualizations across the diverse datasets. A common meaning structure allows for harmonization of separated datasets and affords data sharing. This process enriches the meaning of the data but requires no formalization or standardization of the data themselves to achieve integration (Isaksen et al. 2010). Semantic enrichment is possible through the use of "formal descriptions" and so-called knowledge representation languages, like the Resource Description Framework (RDF) (Sikos 2015). RDF is a way of modeling knowledge and describing semantic content through machine-readable tripartite statements (subject—predicate—object/Graph). RDF statements live within specific formalized "conceptualizations" of knowledge domains or digital ontologies. Digital ontologies give "conceptualization through a data model to a piece of the world" or a specific area of interest (Isaksen 2011; Lukas, Engel, and Mazzucato 2018; Sikos 2015, 2017). Ontologies therefore "provide a shared and common understanding of data" and their processes within a domain (Cripps et al. 2004, 3). The semantic formalization of data has the advantage of creating machine-readable, structured data and opens up the possibility of automatically finding relations between seemingly unrelated data, enhancing the ability of connecting information between data that are otherwise isolated in different domains (data silos) (Cripps et al. 2004; Isaksen 2011; Sikos 2015).

Within the "semantic web" paradigm, the Linked Open Data (LOD) approach has been largely applied as a convenient and pragmatic way of modeling knowledge (e.g., the Open Context platform; Kansa 2012). While this approach is relatively easy to implement and uses different ontological models to produce data interoperability, it does not create a unique model to describe the meaning of the data (Isaksen et al. 2010; Lukas, Engel, and Mazzucato 2018). The LOD standards were applied to the Çatalhöyük dataset for the 2014 implementation of the LAÇ pilot release (Engel and Grossner 2015). This first release entailed a substantial effort: a large amount of Çatalhöyük tabular data stored in the centralized relational database were published in RDF triplestore format and added to the LOD cloud (OpenRDF Sesame; Engel and Grossner 2015, 19). Almost 2.3 million RDF triplets were published to describe approximately 250,000 finds and their contexts (unit/feature/building/space; Engel and Grossner 2015, 18). A web application was created already in 2014 to harmonize and display ÇRP data and queries easily. The spatial component forms a prominent part of that application and represents an essential mechanism of integration. The GIS dataset is therefore fully integrated, and the manipulation of space is an indispensable tool for the comprehension of datasets, their visualization, and presentation. One of the main objectives of the LAÇ has been to make the spatial component explicit and function together with the textual component, especially the data produced within the reflexive framework (Engel and Grossner 2015). In the development of the LAÇ prototype a PostgreSQL/PostGIS database served as an intermediary step to convert and import a version of the ArcGIS geodatabase. A subsequent step will be the integration of the spatial data into a triplestore framework (Engel and Grossner 2015; Lukas, Engel, and Mazzucato 2018).

As Mellaart (1967) observed in the 1960s, however, Çatalhöyük resists all shortcuts. It has become evident while working on the modeling of archaeological knowledge contained in the ÇRP digital dataset that a comprehensive solution capable of embracing all aspects of the ÇRP recording system cannot be achieved and that the ÇRP digital archive cannot be "adequately formalized as a whole" (Lukas, Engel, and Mazzucato 2018). The complexity of the digital material produced at the site (notably the datasets produced within the reflexive framework) and the idiosyncrasies of its recording system require the design of a tailor-made Çatalhöyük semantic model. To represent the different components of the system and multi-strained research properly and to "formalize" and make transparent the reflexive framework within which they were generated, the different parts should be modeled as separate domains and "task ontologies" (Hohmann and Fichtner 2015; Lukas, Engel, and Mazzucato 2018). This new model will build on the existing high-level ontologies created for heritage and archaeological data (e.g., CIDOC CRM and its extensions CRMarchaeo and CRMba), without fully coinciding with them (Lukas, Engel, and Mazzucato 2018). The CIDOC CRM higher-level ontologies offer a complete and structured way of coding archaeological knowledge and have been under experimentation and review since the early 2000s (LeBoeuf et al. 2017). Additionally, the CIDOC CRM offers a wide array of fully compatible extensions (Hiebel, Doerr, and Eide 2015; Nicolucci 2017; Ronzino et al. 2016; Ronzino 2017). Recently, the Ariadne Project elaborated a global reference model using CIDOC CRM as its backbone, both systematizing existing extensions and adding a suite of new ones (Doerr et al. 2015). This modeling approach will be sustained by a robust "semantic-web platform" (Lukas, Engel, and Mazzucato 2018).

Conclusions

As one clear conclusion to the challenges and successes addressed in the development of systems to perpetuate and make accessible ÇRP data, we would like to emphasize the necessity of developing ad hoc systems, tailor-made for the archives of particular sites and their recording systems. Additionally, we can conclude that trying to find a single data structure model to encompass all the domains of research at a site like Çatalhöyük seems a pointless endeavor. On the contrary, it appears that defining dedicated conceptual models (ontologies) for particular parts of the research is both a more efficient and better way of representing and capturing the complexity of data.

We would additionally like to stress the importance of understanding the genesis of datasets, and the way they are embedded in existing environments, to create sound ontology-based applications and, more generally, applications that aim to integrate different datasets. In this respect, individuals who understand archaeological practice and have strong technical backgrounds are best suited for this task.

Understanding the genesis of data and developing sound ontologies make it necessary to fully engage with datasets and processes. Research—not only at Çatalhöyük—does not allow for shortcuts, and publishing the research of archaeological projects requires a demonstration of the complexity of datasets, and perhaps even more so when data derive from a reflexive, multivocal methodology.

Acknowledgments

We are thankful to Ian Hodder and to all the Çatalhöyük Research Project team members. We are additionally thankful to Karl Grossner for his contribution to the early phase of the LAÇ and to Chris Roosevelt and ANAMED for giving us the opportunity of presenting our work at the Spatial Webs symposium.

Works Cited

Atici, Levent, Sarah Whitcher Kansa, Justin Lev-Tov, and Eric C. Kansa. 2013. "Other People's Data: A Demonstration of the Imperative of Publishing Primary Data." *Journal of Archaeological Method and Theory* 20: 663–81.

Barker, Elton, Stefan Bouzarovski, Christopher Pelling, and Leif Isaksen. 2010. "Mapping an Ancient Historian in a Digital Age: The Herodotus Encoded Space-Text-Image Archive (HESTIA)." *Leeds International Classical Studies* 9 (1): 1–24.

Bayliss, Alex, Fiona Brock, Shahina Farid, Ian Hodder, John Southon, and R. E. Taylor. 2015. "Getting to the Bottom of It All: A Bayesian Approach to Dating the Start of Çatalhöyük." *Journal of World Prehistory* 28 (1): 1–26.

Berggren, Åsa, Nicoló Dell'Unto, Maurizio Forte, Scott Haddow, Ian Hodder, Justine Issavi, Nicola Lercari, Camilla Mazzucato, Allison Mickel, and James S. Taylor. 2015. "Revisiting Reflexive Archaeology at Çatalhöyük: Integrating Digital and 3D Technologies at the Trowel's Edge." *Antiquity* 89 (344): 433–48.

Berggren, Åsa, and Bjorn Nilsson. 2014. "Going Back, Looking Forward: Reflexive Archaeology or Reflexive Method?" In *Integrating* Çatalhöyük: *Themes from the 2000–2008 Seasons*, edited by Ian Hodder, 57–70. Çatalhöyük Research Project, vol. 10. London: British Institute at Ankara.

Berggren, Åsa, and Ian Hodder. 2003. "Social Practice, Method, and Some Problems of Field Archaeology." *American Antiquity* 68 (3): 421–34.

Bevan, Andrew. 2015. "The Data Deluge." *Antiquity* 89 (348): 1473–84.

Cessford, Craig. 2005. "Estimating the Neolithic Population of Çatalhöyük." In *Inhabiting Çatalhöyük: Reports from the 1995–99 Seasons*, edited by Ian Hodder, 323–26. London: British Institute at Ankara.

Conolly, James, and Mark Lake. 2006. *Geographical Information Systems in Archaeology*. Cambridge: Cambridge University Press.

Cripps, Paul, Anne Greenhalgh, Dave Fellows, Keith May, and David Robinson. 2004. *Ontological Modelling of the Work of the Centre for Archaeology*. http://old.cidoc-crm.org/docs/Ontological_Modelling_Project_Report_%20Sep2004.pdf.

Dell'Unto, Nicoló, Giacomo Landeschi, Anne-Marie Leander Touati, Matteo Dellepiane, Marco Callieri, and Daniele Ferdani. 2015. "Experiencing Ancient Buildings from a 3D GIS Perspective: A Case Drawn from the Swedish Pompeii Project." *Journal of Archaeological Method and Theory* 23 (1): 73–94.

Doerr, Martin, Maria Theodoridou, Edeltraud Aspöck, and Anja Masur. 2015. "Mapping Archaeological Databases to CIDOC-CRM." In *CAA 2015 Keep the Revolution Going: Proceedings of the 43rd Annual Conference on Computer Applications and Quantitative Methods in Archaeology*, edited by Stefano Campana, Roberto Scopigno, Gabriella Carpentiero, and Marianna Cirillo, 443–51. Oxford: Archaeopress.

Edgeworth, Matt. 2014. "From Spade-Work to Screen-Work: New Forms of Archaeological Discovery in Digital Space." In *Visualization in the Age of Computerization*, edited by Annamaria Carusi, Aud Sissel Hoel, Timothy Webmoor, and Steve Woolgar, 40–58. London: Routledge.

Engel, Claudia, and Karl Grossner. 2015. "Representing the Archaeological Process at Çatalhöyük in a Living Archive." In *Assembling Çatalhöyük*, edited by Ian Hodder and Arkadiusz Marciniak, 13–23. Leeds: Maney Publishing.

Faniel, Ixchel, Eric Kansa, Sarah Whitcher Kansa, Julianna Barrera-Gomez, and Elizabeth Yakel. 2013. "The Challenges of Digging Data: A Study of Context in Archaeological Data Reuse." In *JCDL 2013 Proceedings of the 13th ACM/IEEE-CS Joint Conference on Digital Libraries*, 295–304. New York: ACM.

Faniel, Ixchel M., and Ann Zimmerman. 2011. "Beyond the Data Deluge: A Research Agenda for Large-Scale Data Sharing and Reuse." *International Journal of Digital Curation* 6 (1): 58–69.

Farid, Shahina. 2015. "Proportional Representations: Multiple Voices in Archaeological Interpretation at Çatalhöyük." In *Material Evidence: Learning from Archaeological Practice*, edited by Robert Chapman and Alison Wylie, 59–78. London: Routledge.

Farid, Shahina, and Ian Hodder. 2014. "Excavation, Recording and Sampling Methodologies." In *Çatalhöyük Excavations: The 2000–2008 Seasons*, edited by Ian Hodder, 91–130. Los Angeles: Cotsen Institute of Archaeology Press.

Forte, Maurizio, Nicoló Dell'Unto, Kristina Jonsson, and Nicola Lercari. 2015. "Interpretation Process at Çatalhöyük Using 3D." In *Integrating Çatalhöyük: Themes from the 2000–2008 Seasons*, edited by Ian Hodder, 43–57. Çatalhöyük Research Project, vol. 10. London: British Institute at Ankara.

Hacıgüzeller, Piraye. 2012. "GIS, Critique, Representation and Beyond." *Journal of Social Archaeology* 12 (2): 245–63.

Hacıgüzeller, Piraye. 2017. "Archaeological (Digital) Maps as Performances: Towards Alternative Mappings." *Norwegian Archaeological Review* 50 (2): 149–71.

Hartzler, Bruce. 2009. "Applying New Technologies." In *The Athenian Agora: New Perspectives on an Ancient Site*, edited John McK. Camp and Craig A. Mauzy, 128–37. Mainz am Rhein: Verlag Philipp von Zabern.

Hiebel, Gerald, Martin Doerr, and Øyvind Eide. 2015. "Integration of CIDOC CRM with OGC Standards to Model Spatial Information." In *Across Space and Time: Papers from the 41st Conference on Computer Applications and Quantitative Methods in Archaeology, Perth, 25–28 March 2013*, edited by Arianna Traviglia, 303–10. Amsterdam: Amsterdam University Press.

Hodder, Ian. 1996. "Re-Opening Çatalhöyük." In *On the Surface: Çatalhöyük 1993–95*, edited by Ian Hodder, 1–17. Cambridge: McDonald Institute for Archaeological Research.

———. 1997. "'Always Momentary, Fluid and Flexible': Towards a Reflexive Excavation Methodology." *Antiquity* 71 (273): 691–700.

———, ed. 2000a. *Towards Reflexive Method in Archaeology: The Example at Çatalhöyük.* Cambridge: McDonald Institute for Archaeological Research, University of Cambridge.

———. 2000b. "Where Is Çatalhöyük? Developing a Reflexive Method in Archaeology." In *Towards Reflexive Method in Archaeology: The Example at Çatalhöyük*, edited by Ian Hodder, 101–10. Cambridge: McDonald Institute for Archaeological Research, University of Cambridge.

———. 2003. "Archaeological Reflexivity and the 'Local' Voice." *Anthropological Quarterly* 76 (1): 55–69.

———. 2008. "Multivocality and Social Archaeology." In *Evaluating Multiple Narratives: Beyond Nationalist, Colonialist, Imperialist Archaeologies*, edited by Junko Habu, Clare Fawcett, and John M. Matsunaga, 196–200. New York: Springer.

———. 2014. "Çatalhöyük: The Leopard Changes Its Spots; A Summary of Recent Work." *Anatolian Studies* 64: 1–22.

———. 2015. "Assembling Science at Çatalhöyük: Interdisciplinarity in Theory and Practice." In *Assembling Çatalhöyük*, edited by Ian Hodder and Arkadiusz Marciniak, 7–12. Leeds: Maney Publishing.

Hodder, Ian, Craig Cessford, and Shahina Farid. 2007. "Introduction to Methods and Approach." In *Excavating Çatalhöyük: South, North and KOPAL Area Reports from the 1995–1999 Seasons*, edited by Ian Hodder, 3–24. Cambridge: McDonald Institute for Archaeological Research.

Hohmann, George, and Mark Fichtner. 2015. "Chancen und Herausforderungen in der praktischen Anwendung von Ontologien für das Kulturerbe." In *Digitales Kulturerbe: Bewahrung und Zugänglichkeit in der wissenschaftlichen Praxis*, edited by Caroline Y. Robertson-von Trotha and Ralf H. Schneider, 115–128. Kulturelle Überlieferung – Digital 2. Karlsruhe: KIT Scientific Publishing.

Isaksen, Leif. 2011. "Archaeology and the Semantic Web." PhD diss., University of Southampton.

Isaksen, Leif, Kirk Martinez, Nicholas Gibbins, Graeme Earl, Simon Keay, and Nicholas Gibbons. 2010. "Linking Archaeological Data." In *Making History Interactive: Computer Applications and Quantitative Methods in Archaeology (CAA); Proceedings of the 37th International Conference, Williamsburg, Virginia, United States of America, March 22–26, 2009*, edited by Bernard Frischer, Jane Webb Crawford, and David Koller, 130–36. Oxford: Archaeopress.

Isaksen, Leif, Rainer Simon, Elton T. E. Barker, and Pau de Soto Cañamares. 2014. "Pelagios and the Emerging Graph of Ancient World Data." In *WebSci '14: Proceedings of the 2014 ACM Conference on Web Science*, 197–201. New York: ACM.

Jackson, Sarah E., Christopher F. Motz, and Linda A. Brown. 2016. "Pushing the Paperless Envelope." *Advances in Archaeological Practice* 4 (2): 176–91.

Kansa, Eric. 2012. "Openness and Archaeology's Information Ecosystem." *World Archaeology* 44 (4): 498–520.

Kintigh, Keith W., Jeffrey H. Altschul, Ann P. Kinzig, W. Fredrick Limp, William K. Michener, Jeremy A. Sabloff, Edward J. Hackett, Timothy A. Kohler, Bertram Ludäscher, and Clifford A. Lynch. 2015. "Cultural Dynamics, Deep Time, and Data." *Advances in Archaeological Practice* 3 (1): 1–15.

Landeschi, Giacomo, Nicoló Dell'Unto, Karin Lundqvist, Daniele Ferdani, Danilo Marco Campanaro, and Anne-Marie Leander Touati. 2016. "3D-GIS as a Platform for Visual Analysis: Investigating a Pompeian House." *Journal of Archaeological Science* 65: 103–13.

LeBoeuf, Whitney, Katherine Barghaus, Cassandra Henderson, Kristen Coe, John Fantuzzo, and James Moore. 2017. "The Use of Integrated Data to Inform Quality Pre-K Expansion in Philadelphia." *Research Briefs* 1. https://repository.upenn.edu/pennchild_briefs/1.

Lucas, Gavin. 2012. *Understanding the Archaeological Record*. Cambridge: Cambridge University Press.

Lukas, Dominik, Claudia Engel, and Camilla Mazzucato. 2018. "Towards a Living Archive: Making Multi Layered Research Data and Knowledge Generation Transparent." *Journal of Field Archaeology* 43: 19–30.

Manovich, Lev. 2001. *The Language of New Media*. Cambridge, MA: MIT Press.

Marciniak, Arkadiusz. 2015. "Introduction: Assembling the Archaeological Process at Çatalhöyük." In *Assembling Çatalhöyük*, edited by Ian Hodder and Arkadiusz Marciniak, 1-5. Leeds: Maney Publishing.

Mellaart, James. 1962. "Excavation at Çatal Hüyük: First Preliminary Report, 1961." *Anatolian Studies* 12: 41-65.

———. 1963. "Excavations at Çatal Hüyük, 1962: Second Preliminary Report." *Anatolian Studies* 13: 43–103.

———. 1964. "Excavations at Çatal Hüyük, 1963: Third Preliminary Report." *Anatolian Studies* 14: 39–119.

———. 1966. "Excavations at Çatal Hüyük, 1965: Forth Preliminary Report." *Anatolian Studies* 16, 165–191.

———. 1967. *Çatal Hüyük: A Neolithic Town in Anatolia*. New York: McGraw-Hill.

Mickel, Allison. 2015. "Reasons for Redundancy in Reflexivity: The Role of Diaries in Archaeological Epistemology." *Journal of Field Archaeology* 40 (3): 300–309.

Morgan, Colleen Leah. 2012. "Emancipatory Digital Archaeology." PhD diss., University of California, Berkeley.

Morgan, Colleen, and Holly Wright. 2018. "Pencils and Pixels: Drawing and Digital Media in Archaeological Field Recording." *Journal of Field Archaeology* 43 (2): 136–51.

Nicolucci, Franco. 2017. "Documenting Archaeological Science with CIDOC CRM." *International Journal on Digital Libraries* 18 (3): 223–31.

Olson, Brandon R, Ryan A Placchetti, Jamie Quartermaine, and Ann E Killebrew. 2013. "The Tel Akko Total Archaeology Project (Akko, Israel): Assessing the Suitability of Multi-Scale 3D Field Recording in Archaeology." *Journal of Field Archaeology* 38 (3): 244–62.

Opitz, Rachel. 2014. "Three Dimensional Field Recording in Archaeology: An Example from Gabii." In *Visions of Substance: 3D Imaging in Mediterranean Archaeology*, edited by Brandon Olson and William Caraher, 64–73. The Digital Press at the University of North Dakota. https://thedigitalpress.org/portfolio/visions-of-substance-3d-imaging-in-mediterranean-archaeology/.

Richards-Rissetto, Heather. 2017. "What Can GIS + 3D Mean for Landscape Archaeology?" *Journal of Archaeological Science* 84: 10–21.

Ridge, Mia, and Richard May. 2004. "Database & IT Developments." In *Çatalhöyük 2004 Archive Report*. http://www.catalhoyuk.com/archive_reports/2004/ar04_32.html.

Ridge, Mia, Richard May, and Dave Mackie. 2005. "Çatalhöyük IT Team Archive Report 2005." In *Çatalhöyük 2005 Archive Report*. http://www.catalhoyuk.com/archive_reports/2005/ar05_37.html.

Ronzino, Paola. 2017. "Harmonizing the CRMba and CRMarchaeo Models." *International Journal on Digital Libraries* 18 (4): 253–61.

Ronzino, Paola, Franco Niccolucci, Achille Felicetti, and Martin Doerr. 2016. "CRMba a CRM Extension for the Documentation of Standing Buildings." *International Journal on Digital Libraries* 17 (1): 71–78.

Roosevelt, Christopher H., Peter Cobb, Emanuel Moss, Brandon R. Olson, and Sinan Ünlüsoy. 2015. "Excavation Is ~~Destruction~~ Digitization: Advances in Archaeological Practice." *Journal of Field Archaeology* 40 (3): 325–46.

Shanks, Michael, and Timothy Webmoor. 2012. "A Political Economy of Visual Media in Archaeology." In *Re-Presenting the Past: Archaeology through Text and Image*, edited by Sheila Bonde and Stephen Houston, 142–82. Oxford: Oxbow Books.

Sikos, Leslie. 2015. *Mastering Structured Data on the Semantic Web: From HTML5 Microdata to Linked Open Data*. New York: Apress.

Taylor, James, Amy Bogaard, Tristan Carter, Michael Charles, Scott Haddow, Christopher J. Knüsel, Camilla Mazzucato, et al. 2015. "'Up in Flames': A Visual Exploration of a Burnt Building at Çatalhöyük in GIS." In *Assembling Çatalhöyük*, edited by Ian Hodder and Arkadiusz Marciniak, 127–49. Leeds: Maney Publishing.

Taylor, James, Justine Issavi, Åsa Berggren, Dominik Lukas, Camilla Mazzucato, Burcu Tung, and Nicoló Dell'Unto. 2018. "'The Rise of the Machine': The Impact of Digital Tablet Recording in the Field at Çatalhöyük." *Internet Archaeology* 47. https://doi.org/10.11141/ia.47.1

Spatializing the Sagalassos Integrated Information System

Anuja Dangol

Piraye Hacıgüzeller

Jeroen Poblome

Introduction

Archaeology is intrinsically spatial in the sense that archaeological phenomena almost exclusively have a spatial component. It is therefore hardly surprising that tools (be these conceptual, analog, digital, or hybrid) that assist in collecting, storing, querying, processing, displaying, and/or studying geospatial data and information will always have a key role in archaeology. Geographical Information Systems (GIS) in this context have a notable place given that the technology revolutionized spatial archaeology and related analyses. It was rapidly adopted in the early 1990s and is still widely used, representing possibilities for innovative and at times extremely prolific archaeological research workflows (Lock and Pouncett 2017; Verhagen 2018).

The real boom of geospatial technologies and practices in archaeology came in the 2000s with advances in remote sensing applications (such as airborne LiDAR and satellite imaging), open-source and free GIS software packages, spatial analysis, geographic visualization, 3-D modeling, geoparsing and geographic text analysis, and linked open geodata. The ubiquity of these technologies and practices has increased even more rapidly in recent years (e.g., Gupta and Devillers 2017; Simon et al. 2017; Dell'Unto et al. 2016; Gregory et al. 2015; Hritz 2014; Chase et al. 2012). Additionally, the internet spread globally in the 2000s, particularly Web 2.0—which can be defined as participatory internet (Gartner 2009; Dodge and Kitchin 2013)—and the mobile web (Peterson 2008), which have made the practices of circulating, accessing, editing, and consulting geospatial information widespread. This rapid and abundant influx of new geospatial tools and practices to archaeology is a significant phenomenon that needs close attention, given that "new tools make possible the production of entirely new sorts of data, information, interpretation and, ultimately, archaeology" (Beale and Reilly 2017). As a result, we seem to be at the right moment to discuss archaeological geospatial technologies, their effects, and the relations in which they are embedded.

In this chapter we aim to present the Sagalassos Integrated Information System (SIIS), a digital database designed for the Sagalassos Archaeological Research Project (henceforth Sagalassos Project) and based on PostgreSQL, an open-source and free object-relational database management system. The chapter not only refers to the technological specifics of the SIIS but also narrates challenges encountered during its expansion in recent years, including the creation of its geospatial component. It concludes with a discussion on how the future of data management in the Sagalassos Project is currently assessed.

The Archaeological Site of Sagalassos and the SIIS

Sagalassos is an archaeological site located in southwest Turkey, near Ağlasun (Burdur). Roughly one hundred kilometers from Antalya, the site is situated in the Taurus mountains at ca. 1,450–1,600 m above sea level. Evidence for human activity in the Sagalassos study region ranges from Middle Paleolithic times onward. The region holds its most important archaeological potential for the Late Neolithic/Early Chalcolithic, Early Bronze Age, and Classical periods. The settlement of Sagalassos originated in Late Achaemenid times but grew into a polis by the Middle Hellenistic period. The town saw its heyday during the first to third centuries CE, when it identified itself as a true Roman city. At that time, it became a regional center of the imperial cult and quickly expanded with new building projects and intense economic activities. Ancient DNA analysis indicates that the local community survived the waves of the Justinianic plague fairly intact, but the urban archaeological record does show indications of decline by the second half of the sixth century. A major earthquake around the middle of the seventh century sealed the fate of most of the urban monumental architecture, but once again the community remained relatively intact. In Byzantine times, the settlement regained some potential and maintained some central governance functions. By the mid-thirteenth century, finally, organized community life came to an end at Sagalassos (cf. Daems and Poblome 2017; Poblome, Talloen and Kaptijn 2017; Talloen and Poblome 2016).

Systematic archaeological fieldwork has been ongoing at Sagalassos since 1986, coordinated by the University of Leuven since 1990. A large amount of information with considerable variety has been collected at the site during this fieldwork. From 1993 onward, various interdisciplinary survey campaigns in the region resulted in the addition of considerable quantities of geospatial and associated information. Information on objects, architectural fragments, stratigraphic units, and more was recorded largely in analog form in the early stages of the Sagalassos Project. From 2006 onward, the project started to invest in designing and developing a digital relational database, what became the SIIS,[1] hosted during winter months on University of Leuven servers and during summer seasons on the excavation house servers at Sagalassos.

1 Importantly, the SIIS has always existed beside a set of other project databases owing to the project's interdisciplinary structure from its onset. These other databases had been developed by participating disciplines and individual researchers, at all levels of complexity. Moreover, a photographic and cartographic archive was managed centrally by the Sagalassos Project without being a part of the SIIS.

Until the early 2000s, geospatial data (specifically, three-dimensional coordinate data) collected at Sagalassos had either been kept in tabular format or were entered into corresponding analog forms during fieldwork. From the 2000s onward, the project incorporated GIS in its research workflow, and the technology became the project's means to store, manage, and display geospatial data. Since the SIIS included no geospatial component at first, even after its implementation in the field, GIS software remained the primary means to deal with geospatial information collected at Sagalassos. This situation resulted in the need to deal with two data management systems to manage the large volume of information available: a GIS database with very limited attribute data and a relational database with an extensive amount of attribute data. Both databases represented official project databases, were managed centrally by the team, and included all basic administrative and archaeological information resulting from every stage of fieldwork and depot registration procedures. Both were mostly geared toward the registration of new data generated through fieldwork.

The need to combine the geospatial and attribute databases at Sagalassos became clearer over the years for three main reasons. Firstly, the motivation to combine these databases emerged from the need to carry out spatiotemporal queries that relied on absolute locations of database entries, such as stratigraphic units (defined as "loci" in the Sagalassos Project), objects, and architectural fragments. Secondly, it became clear that the existing attribute database system managed through the ArchaeoNotes application was limited in scope: it did not cover all aspects of the attribute data created in Sagalassos, and it needed to be expanded (Fig. 1).[2] It was a result of this particular limitation that simple excel sheets were used extensively at times for data collection in the field, for instance to register details of architectural description or to enter information for material category studies. Thirdly, the combination of the two databases was expected to allow the visualization of information from the site in its inherent complexity by archaeologists and others, especially during the process of fieldwork and depot studies.[3] This final point was deemed important for making archaeological practices at Sagalassos more reflexive.

---

2 The ArchaeoNotes application was tailor-developed in Java for managing and populating the SIIS in its early years. Besides its use in pre- and post-excavation stages, it was also installed on rugged tablets and laptops, mainly to be employed for data collection during fieldwork. Data in ArchaeoNotes could be exported as PostgreSQL tables. Figure 1 shows the relatively simple SIIS database structure during its use with ArchaeoNotes. The database included information on loci, architectural fragments, different types of artifacts, and samples. After encountering copyright issues with the application and realizing it was unfit for the prospective expansion of the SIIS, the decision was taken to cease use of ArchaeoNotes at Sagalassos from 2018 onward and develop a new application for the expanded and spatialized SIIS, as explained in this chapter.

3 It needs to be clarified here that the generation of new data in the field is physically separated from the simultaneously ongoing find study processes at Sagalassos: while the first happens on the site in the mountains, the latter takes place in the excavation house in the valley. Yet, all researchers involved find it very important to share new insights as they go along in their part of the archaeological process.

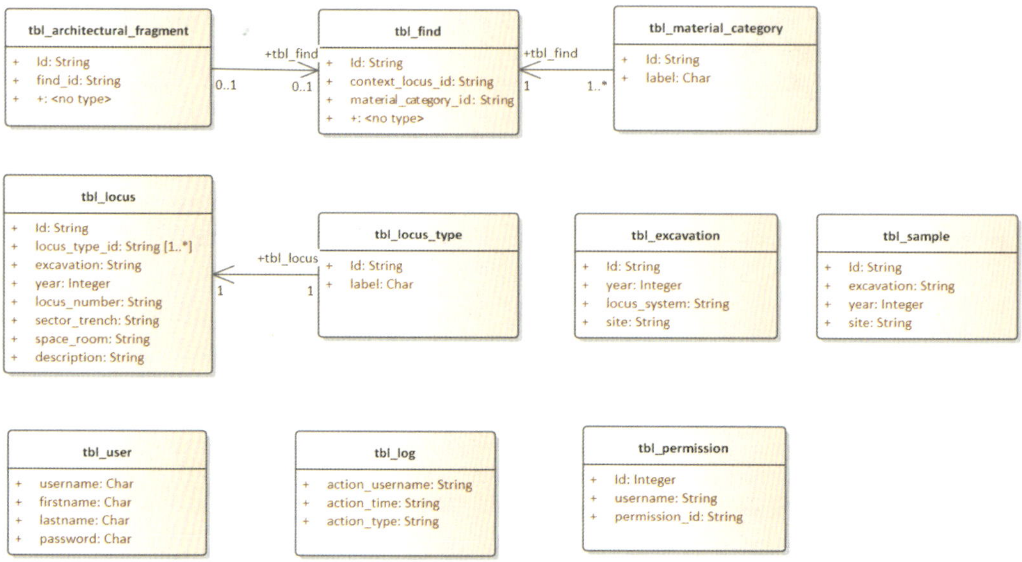

Fig. 1. The relatively simple database structure of the SIIS, prior to its expansion.

Expanding and Spatializing the SIIS

The process of spatializing the SIIS started in 2016 by screening the available digital geospatial datasets collected by the project. The ultimate aim was to select the relevant datasets to be entered into the upgraded version of the SIIS. Throughout the screening, it became clear that the size of Sagalassos Project geospatial information was huge (around 60 GB), including vector data, raster data, and three-dimensional coordinate data in tabular format. In terms of content, the datasets were heterogeneous, including, for example, data on boundaries and locations of geographic entities, land use, and three-dimensional models. During the screening, the spatial datasets were classified on the basis of whether a certain dataset should be migrated into the SIIS or not. Possible answers included "probably yes," "probably no," "yes," and "no." After the first and second screening of geospatial datasets, a final selection was made of datasets to be migrated into the spatialized SIIS. Subsequently, a standard metadata template was implemented to document the information that would be contained in the new SIIS. INSPIRE guidelines were taken as a reference for making the metadata template at this stage (European Commission 2019).

Three-dimensional coordinate information on loci, retrieved during the screening process, was added to the loci attribute database table using Structured Query Language (SQL). Subsequently sample queries were collected from project team members in preparation for designing the new relational database model as well as for assessing how comprehensive the new SIIS would have to be. These queries were classified as geospatial or attribute-based depending on whether one would have to rely on geospatial information to answer them. Most of the queries could be answered through the attribute component of the prospective SIIS structure. For example, one such potential query is "list all stratigraphic units with abundant traces of ceramic production waste at site X." This is a spatial query because it involves a location, that is, "site X." However, it is not a geospatial query because the absolute location of "site X" is not required to answer the question; the query therefore qualifies as an attribute-based query. The prospective geospatial component of the new SIIS, on the other hand, would be involved in providing answers to some unique queries regarding the absolute locations of places and finds, such as "where was Byzantine pottery found at Sagalassos?" The significant added value of integrating coordinate information into the SIIS would also include cartographic data visualization and spatial analysis.

Sample queries were also used to conduct tests on the existing SIIS. Quantum GIS (QGIS) software was employed in this testing process. Specifically, QGIS was used to connect to the existing SIIS through a database manager plugin (DB Manager), and outputs were displayed as GIS vector files (shapefiles) in the map canvas. Executing these queries on the existing SIIS showed clearly that on-site attribute and geospatial data collection routines needed improvement and guidelines. This is because different team members entered the same information into the database differently (e.g., "4th century AD," "4th cent. AD," or "fourth cent AD"). It became clear, therefore, that validation rules needed to be applied for data entry in the new SIIS to control the type of data or the values users entered.

After getting a broad idea about how the new SIIS would have to be structured through these sample queries, the process of building the new database started. The first step was strategic analysis with team members and conceptual modeling. At this stage, what needed to be included in the expanded SIIS and what not were discussed extensively within the team. Several versions of database structure were proposed, and versions were discussed iteratively and reworked. In this discussion process, Unified Modeling Language (UML)—a standard language for visualizing, modeling, and documenting information systems—was employed. Using UML allowed better communication of ideas between team members and more effective system design. Modelio, an open-source UML tool (modelio.org/), was used in the modeling process. Figures 2 and 3 illustrate the final UML diagrams of the new data model. Compared to the old data model shown in Figure 1, major changes were made to both tables and relationships that involved loci and artifactual finds. After completing the diagram of the new database structure, the next step was to create data tables and fill them with sample data to double-check whether the data collected at Sagalassos corresponded properly to the designed database tables.

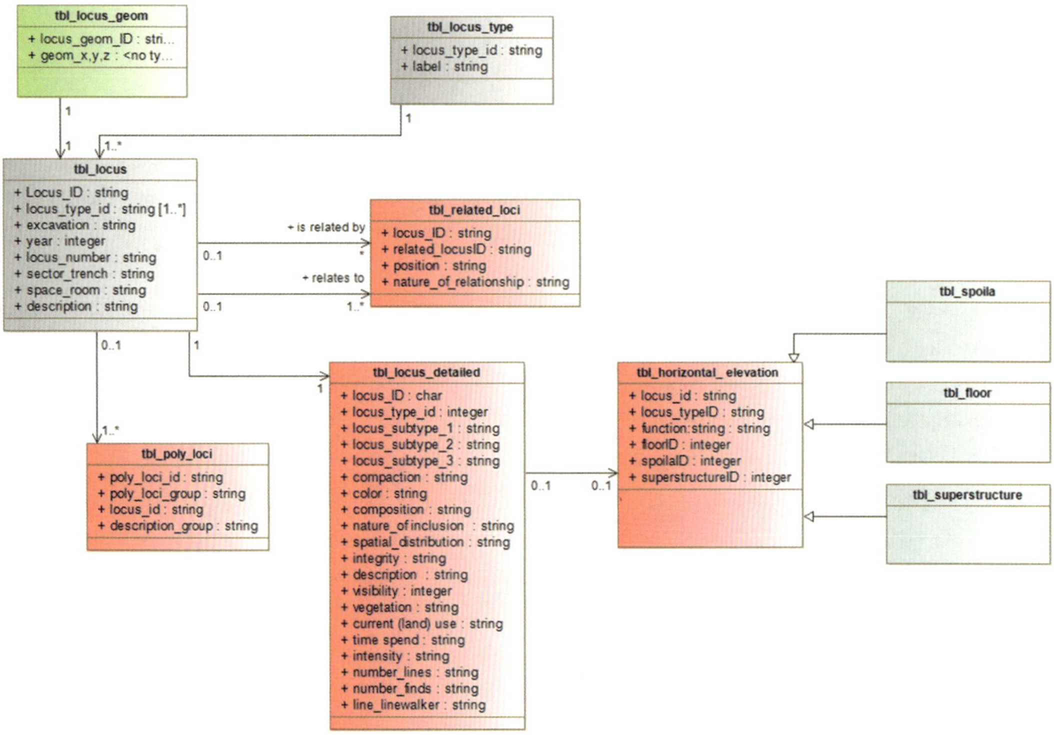

Fig. 2. UML diagram of loci in the new SIIS data model.

A Custom-Built SIIS Application

Presently, the Sagalassos Project is at the application development stage for its new SIIS. The functional requirements of the stand-alone SIIS application are being determined through consultations with team members. Given that the internet connection at Sagalassos can at times be sketchy, the application will have to work offline to collect information. Synchronization at a later stage will be possible when an internet connection is available. The application will also have to allow collecting geospatial vector data in the format of points, polylines, or polygons directly through an internal or external GPS device. Since rugged tablets and laptops running the Windows 8 operating system are used on-site, compatibility will have to be a priority in the short term, although the plan is to render the application platform-independent in the near future.

The Sagalassos Project's preference at the application development stage has also been free and open-source software. To be specific, for creating the graphical user interface (GUI) of the application, which constitutes forms and tabs for collecting different types of attribute information (e.g., on loci, finds, samples) (Fig. 4), PyQT is used. PyQT is a set of Python bindings for the QT cross-platform application framework. The code generated through PyQT is edited in the Python integrated development

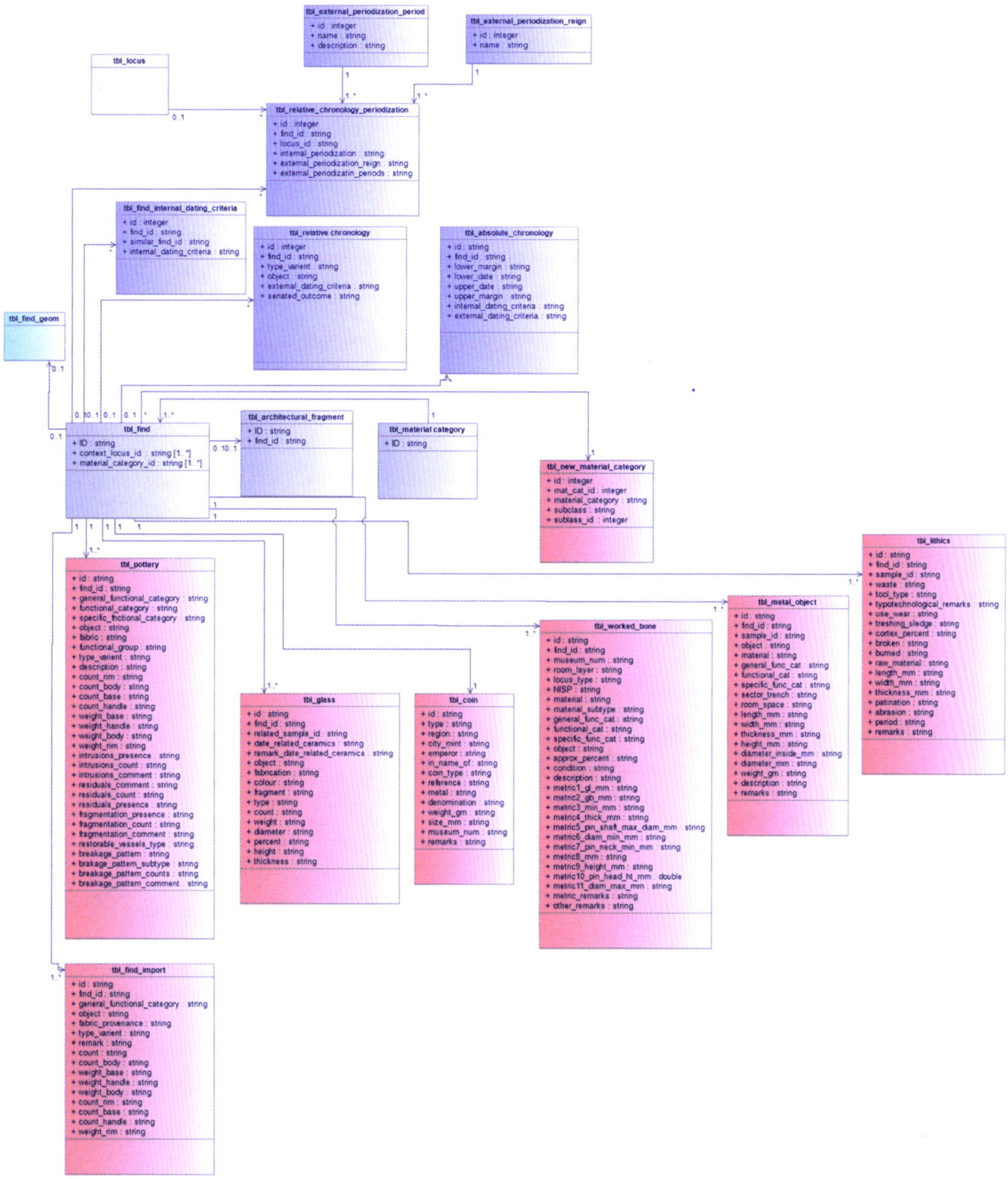

Fig. 3. UML diagram of finds in the new SIIS data model.

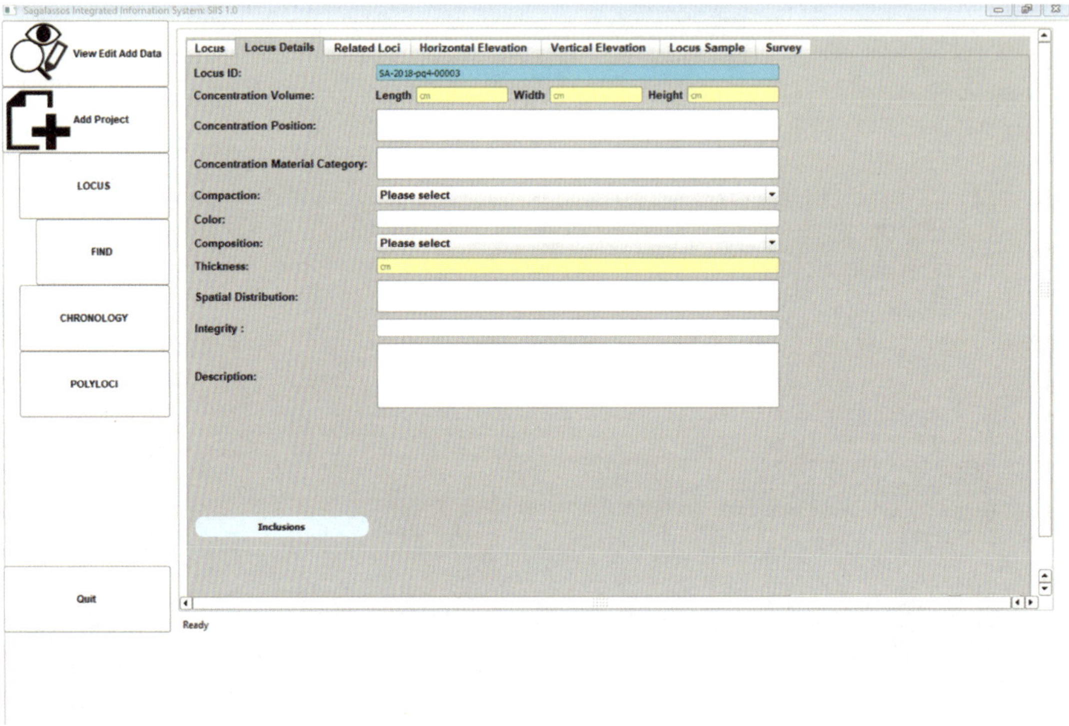

Fig. 4. Screenshot of the GUI of the new SIIS application, showing the "Locus Details" tab.

environment (IDE). The Python for QGIS (PyQGIS) and QGIS application programming interface (API) are planned to be used for developing the geospatial component of the application. This component will be employed for collecting field data by drawing points, lines, and polygons mainly assisted by georeferenced maps. The vector data created through the use of the geospatial component of the application will be stored in PostGIS, a spatial database extension of PostgreSQL. The QGIS DB Manager plug-in is planned to be employed in the application for connecting to the PostgreSQL database, executing queries, and visualizing results. A SpatiaLite database management system will be used for working offline with the geospatial information and later synchronizing modifications with the PostgreSQL database on the SIIS server.

The Future of the SIIS

As discussed above, considerable progress has been made by the Sagalassos Project in recent years in terms of expanding and spatializing the SIIS. Yet, there is certainly more work ahead, particularly in three domains. The first of these concerns the backlog of Sagalassos Project digital geospatial information that is in a vector (and almost exclusively shapefile) format. These data, generated in earlier GIS applications, need to be curated and imported to the SIIS using PostGIS software.

The second domain involves issues concerning metadata, linked data, data archiving, and online data publishing. A data layer that contains metadata in CIDOC CRM standard ontology (Doerr, Ore, and Stead 2007) needs to be created from the new version of the SIIS to render the SIIS database structure semantically explicit. This will serve as an investment in making Sagalassos data semantically retrievable once it is linked to other external datasets in the future, potentially through common references to places (Isaksen et al. 2014), historical events (Shaw, Troncy, and Hardman 2009), or people. There is also a strong case to be made in favor of archiving (a subset of) Sagalassos data in existing online repositories in a machine-readable way (see Richards 2006, 971). Publishing Sagalassos data online through a web application is also one of the primary future aims of the Sagalassos Project, as is providing an API for users who want to interact with the web application and data in their own ways. Two types of equally viable workflows are available in this context and need to be decided on: producing a web application tailor-made for Sagalassos from scratch or altering the code of an existing free and open-source application.

Thirdly and finally, the digital workflow and related tools presented in this article for collecting, managing, and visualizing the attribute and geospatial information at Sagalassos would most likely be useful for other archaeological projects. In this regard, the ability to highlight the presence of new code and software, as well as their updates, in a community of digitally minded archaeologists would avoid reinventing the wheel. For this reason, the Sagalassos Project would like to contribute to efforts to build infrastructures for archaeology that facilitate the sharing of codes and experiences in digital workflows and practices in a way that is free of technical jargon. Many development platforms are available to make such collaborative development practices possible—GitHub perhaps the best known of them. The medium to share code is obviously not the issue here. There is, however, still a considerable need to emphasize the necessity for and value of such digital collaborative efforts in archaeology (e.g., Morgan and Eve 2012). After all, archaeologists' resources for investing in computer science expertise are limited, and, thus, it is to all archaeologists' benefit to pool resources, at least digitally, for the betterment of our data management and sharing potentials.

Acknowledgments

This research was supported by the Research Foundation - Flanders (FWO), the Belgian Programme on Interuniversity Poles of Attraction, and the Research Fund of the University of Leuven.

Works Cited

Beale, Gareth, and Paul Reilly. 2017. "After Virtual Archaeology: Rethinking Archaeological Approaches to the Adoption of Digital Technology." *Internet Archaeology* 44. https://doi.org/10.11141/ia.44.1.

Chase, Arlen F., Diane Z. Chase, Christopher T. Fisher, Stephen J. Leisz, and John F. Weishampel. 2012. "Geospatial Revolution and Remote Sensing LiDAR in Mesoamerican Archaeology." *Proceedings of the National Academy of Sciences* 109 (32): 12916–21.

Daems, Dries, and Jeroen Poblome. 2017. "The Pottery of Late Achaemenid Sagalassos: An Overview." *HEROM: Journal on Hellenistic and Roman Material Culture* 6 (1): 49–62.

Dell'Unto, Nicoló, Giacomo Landeschi, Anne-Marie Leander Touati, Matteo Dellepiane, Marco Callieri, and Daniele Ferdani. 2016. "Experiencing Ancient Buildings from a 3D GIS Perspective: A Case Drawn from the Swedish Pompeii Project." *Journal of Archaeological Method and Theory* 23 (1): 73–94. https://doi.org/10.1007/s10816-014-9226-7.

Dodge, Martin, and Rob Kitchin. 2013. "Crowdsourced Cartography: Mapping Experience and Knowledge." *Environment and Planning A: Economy and Space* 45 (1): 19–36. https://doi.org/10.1068/a44484.

Doerr, Martin, Christian-Emil Ore, and Stephen Stead. 2007. "The CIDOC Conceptual Reference Model: A New Standard for Knowledge Sharing." In *Tutorials, Posters, Panels and Industrial Contributions at the 26th International Conference on Conceptual Modeling*, edited by J. Grundy, S. Hartmann, A. H. F. Laender, L. Maciaszek, and J. F. Roddick, 51–56. Canberra: Australian Computer Society.

European Commission. 2019. "Inspire Knowledge Base: Infrastructure for Spatial Information in Europe." https://inspire.ec.europa.eu/.

Gartner, Georg. 2009. "Web Mapping 2.0." In *Rethinking Maps*, edited by Martin Dodge, Rob Kitchin, and Chris Perkins, 68–82. London: Routledge.

Gregory, Ian, Christopher Donaldson, Patricia Murrieta-Flores, and Paul Rayson. 2015. "Geoparsing, GIS, and Textual Analysis: Current Developments in Spatial Humanities Research." *International Journal of Humanities and Arts Computing* 9 (1): 1–14. https://doi.org/10.3366/ijhac.2015.0135.

Gupta, Neha, and Rodolphe Devillers. 2017. "Geographic Visualization in Archaeology." *Journal of Archaeological Method Theory* 24 (3): 852–85.

Hritz, Carrie. 2014. "Contributions of GIS and Satellite-Based Remote Sensing to Landscape Archaeology in the Middle East." *Journal of Archaeological Research* 22 (3): 229–76. https://doi.org/10.1007/s10814-013-9072-2.

Isaksen, Leif, Rainer Simon, Elton T. E. Barker, and Pau de Soto Cañamares. 2014. "Pelagios and the Emerging Graph of Ancient World Data." In *WebSci '14: Proceedings of the 2014 ACM Conference on Web Science*, 197–201. New York: ACM.

Lock, Gary, and John Pouncett. 2017. "Spatial Thinking in Archaeology: Is GIS the Answer?" *Journal of Archaeological Science* 84: 129–35. https://doi.org/10.1016/j.jas.2017.06.002.

Modeliosoft. 2019. "Modelio: The Open Source Modeling Environment." https://www.modelio.org/.

Morgan, Colleen, and Stuart Eve, 2012. "DIY and Digital Archaeology: What Are You Doing to Participate?" *World Archaeology* 44 (4):521–37. doi: 10.1080/00438243.2012.741810.

Peterson, Michael P. 2008. "Trends in Internet and Ubiquitous Cartography." *Cartographic Perspectives* 61: 36–49.

Poblome, Jeroen, Peter Talloen, and Eva Kaptijn. 2017. "Byzantine Sagalassos." In *The Archaeology of Byzantine Anatolia: From Late Antiquity to the Coming of the Turks*, edited by Philipp Niewöhner, 302–11. Oxford: Oxford University Press.

Richards, Julian D. 2006. "Archaeology, E-publication and the Semantic Web." *Antiquity* 80 (310): 970–79.

Shaw, Ryan, Raphaël Troncy, and Lynda Hardman. 2009. "LODE: Linking Open Descriptions of Events." In *Fourth Annual Asian Semantic Web Conference (ASWC '09)*, 153–67. Berlin: Springer.

Simon, Ranier, Elton Barker, Leif Isaksen, and Pau de Soto Cañamares. 2017. "Linked Data Annotation Without the Pointy Brackets: Introducing Recogito 2." *Journal of Map & Geography Libraries* 13 (1): 111–32. https://doi.org/10.1080/15420353.2017.1307303.

Talloen, Peter, and Jeroen Poblome. 2016. "The 2014 and 2015 Control Excavations on and around the Upper Agora of Sagalassos: The Structural Remains and General Phasing." *Anatolica* 42: 111–50.

Verhagen, Philip. 2018. "Spatial Analysis in Archaeology: Moving into New Territories." In *Digital Geoarchaeology: New Techniques for Interdisciplinary Human-Environmental Research*, edited by Christoph Siart, Markus Forbriger, and Olaf Bubenzer, 11–25. Cham: Springer International Publishing.

Mapping and Analyzing the Pergamon Micro-Region: An Example of Reproducible Research

Daniel Knitter

Bernhard Ludwig

Introduction

Replication, the ultimate standard to judge scientific claims, is difficult to achieve in archaeological practice. Every site is unique and the practice of excavation or collection of material during surveys causes irrevocable change, preventing independent replication of original conditions. Rigorous and maximal comprehensive documentation is still the only way to ensure that scientific standards are met and findings are sound. Today, the integration of computers and computer applications into workflows results in advantages including (1) that data acquisition and documentation rates increase, as does their standardization, and (2) that analyses based on these data are reproducible. In this chapter we present the example of one approach to create a reproducible study, contributing to the growing body of literature on reproducible research in archaeology (e.g., Marwick 2017). We argue that it is high time for archaeologists to learn new computational habits to create reproducible studies and, in this process, to produce results and conclusions whose production processes are transparent and can be taken up, modified, or extended by other researchers.

Reproducible Research

Reproducible research practice is the attainable minimum standard for assessing the value of scientific claims, especially when replication is not feasible or possible (Peng 2011, 1226). It calls for the publication of not only the data but also the computer code used to analyze the data (Peng 2011, 1226). Data and code should be published in research data repositories, like Zenodo,[1] in non-proprietary formats, and using container-frameworks

1 Zenodo (zenodo.org) is a research data repository, created in the course of the OpenAIRE project that

that allow reproduction of analyses irrespective of software versions or operating systems.

Gentleman and Lang (2007, 4) proposed the concept of a research compendium, that is, a container to associate a dynamic document with external data and the auxiliary software needed to generate the paper's content. The general idea of a dynamic document goes back to Knuth (1984), who proposed the concept of literate programming. According to this, a dynamic document interweaves the textual description of the problem together with the computer code that processes that data (Gentleman and Lang 2007, 4). This chapter was created from a dynamic document that is distributed as a research compendium (Fig. 1). We use different auxiliary software to create the final document; the code is written in R (R Core Team 2017) using supplementary libraries packaged within our research compendium. For text, code, data, and necessary software, we created a docker container. Together with instructions on how to use it, it is published at zenodo.org under the DOI 10.5281/zenodo.1161229, allowing interested readers to reproduce our analysis.

Specifically, we employ the following open-source software tools and libraries:

- GDAL (Geospatial Data Abstraction Library; GDAL Development Team 2017): to read/write and pre-process geospatial data
- R (R Core Team 2017) with many contributing packages[2]: to perform the analysis
- GRASS GIS (GRASS Development Team 2017): to calculate specific geomorphometric parameters
- pandoc (version 2.0.6, pandoc.org): to render our literate programming document
- git (version 2.7.4, git-scm.com): to enable version-controlled, collaborative work on the text and code
- docker (version 1.13.1, docker.com): to create a container of the presented study so every interested reader can investigate and reproduce our analysis

The number of reproducible studies in archaeology is still negligible. To change this situation, the editorial standards of archaeological publication need to be revised. The technical prerequisites already exist, as this article demonstrates, but without certain incentives they will not be used (Faris et al. 2011). Another problem is the small number of archaeologists who employ the concept of reproducibility in their daily work, because the concept plays only a subordinate role in humanities-based archaeological

hosts any kind of research in an infrastructure provided by CERN. It helps researchers receive credit for publishing their data by making them citable. Zenodo is funded by the European Commission and provides an alternative to commercial services like figshare (figshare.com).

2 The contributing packages are mentioned at the specific steps where we used them.

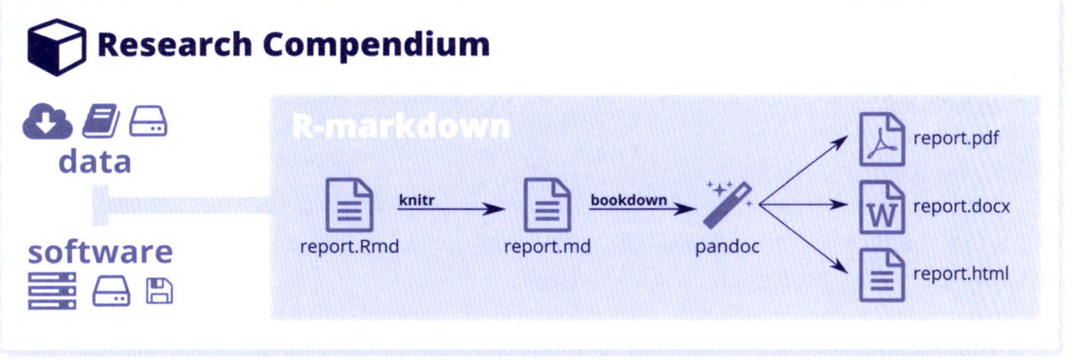

Fig. 1. The structure of the research compendium of this study. Data and software are provided alongside a document that follows the guidelines of literate programming; that is, it combines thematic text and computer code and is published at a research data repository (doi:10.5281/zenodo.1161229).

training. The concept originates from the natural sciences and computer science and is only slowly entering other fields. The proportion of archaeologists with sufficient computer skills to carry out reproducible research is, accordingly, relatively low. Traditional researchers have great expertise in their special disciplines but typically do not focus on the computational details that enable the creation of respective results. A *modern* researcher, hence, needs expanded expertise not only in her discipline, but also in a second field, namely that of computing skills and reproducibility. This idea was last presented by Marwick (2017) in the context of reproducibility in archaeology and T and Pi-shaped researchers (Fig. 2). Thanks to well-documented and easily accessible software tools and a vibrant community of researchers that develop new computational methods, it is easier than ever before to become a Pi-shaped researcher.

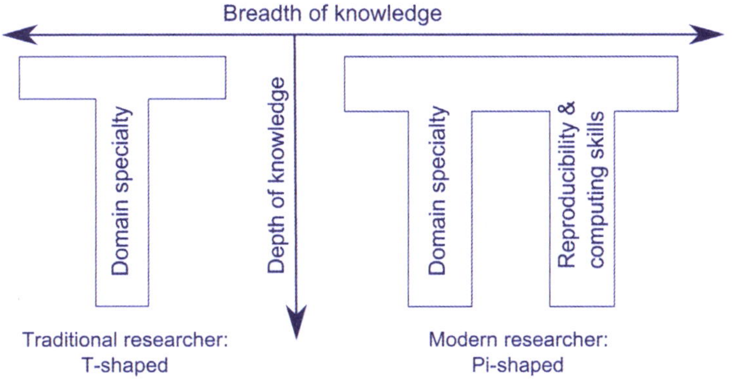

Fig. 2. Because it is grounded in two different subject areas, modern research practice can be imagined as Pi-shaped, combining discipline-specific expertise and computational skills. Future archaeologists should be trained as Pi-shaped researchers instead of T-shaped, that is, with strong expertise in one specific field only (Marwick 2017, 445).

Interaction Networks within the Pergamon Micro-Region

The aim of our case study and published research compendium is to illustrate a reproducible workflow on real-world material, addressing a classic question of archaeological significance. The applied methods, results, and discussion can (and should) be extended in various ways and supplemented with further data. Such next steps remain out of the scope of this study, however, and alternatively we refer interested readers to the work of Pirson and Zimmermann (2014), Pirson (2017), Sommerey (2008), and Zimmermann, Matthaei, and Ateş (2015).

The ancient city of Pergamon (modern Bergama) is located near today's western coast of Turkey, about thirty kilometers inland and on the northern edge of the Kaikos River (modern Bakırçay) Valley. The valley is surrounded by the Pindasos (modern Madra Dağ) Mountains (1,243 m) on the north, the Aspordenon (modern Yunt Dağ) Mountains (782 m) on the south, and the Kara Dağ Mountains (754 m) on the Kane Peninsula to the west (Fig. 3). The western part of the lower Kaikos Valley, between Pergamon and the Aegean coast, is the micro-region of interest in this study. The Attalid Empire of Pergamon existed for several centuries and shaped the history of western Asia Minor (Radt 2016). As the seat of a Hellenistic dynasty, it was vital for Pergamon to have complete control of its immediate territory. It afforded access to essential resources and to major communication routes, for example between the city of Pergamon and its harbor city Elaia (Fig. 3). We hypothesize that to ensure control within this micro-region, an interaction network, established using a network of fortifications, was important (Williamson 2016). Our conception of the term interaction refers to Nakoinz (2013), who developed a concept of interaction with the focus on spatial interaction models and examined definitions, influencing factors, and interaction dimensions.

The term micro-region was established by Horden and Purcell (2000) in historical and classical studies. A micro-region is not primarily defined geographically but as an interaction space between humans and the natural environment. Its extent can therefore vary according to the level of interaction (e.g., economic, military, sociocultural) and is subject to diachronic changes (Schuler 2016). Contrary to the hierarchical concept of the city and its surroundings, we understand Pergamon to be part of a complementary spatial system: the Pergamon Micro-Region.

Until now, only a few computer-based, geospatially quantitative analyses have been conducted on the Pergamon Micro-Region. One example is Williamson's (2014, 2016) study using Geographic Information Systems (GIS) for visibility analyses to explore the interdependencies between mythology (sanctuaries, ritual spaces), landscape, power, and territory. Several archaeological surveys with temporal foci ranging from prehistoric to Roman times have been conducted since 2006 under the direction of the Pergamon Excavation Project of the German Archaeological Institute (DAI). However, the evaluation of collected data has not yet gone beyond the level of mapping. Studies on interaction are still on the level of heuristics (e.g., Knitter et al. 2013), and GIS-based

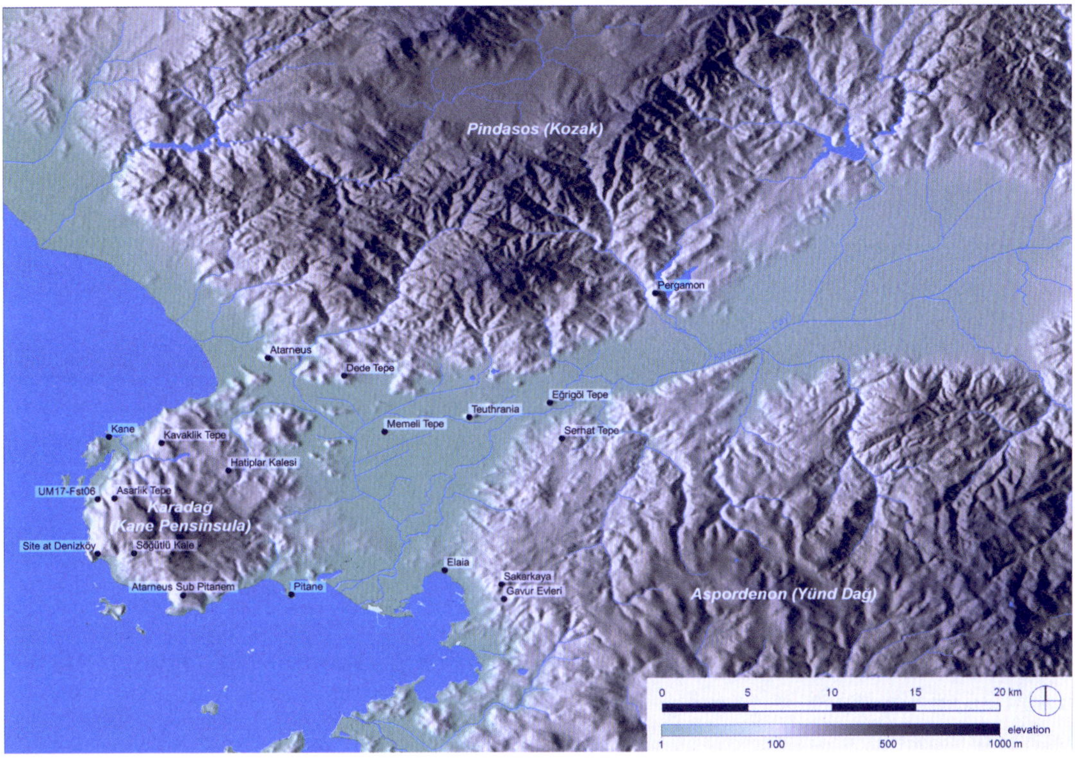

Fig. 3. Map of the lower Kaikos River Valley with surrounding mountains. Elevation information is based on EU-DEM (EU Copernicus Programme 2016). The waterways and coastline are based on OpenStreetMap (2018) and data from the Copernicus Programme (2016).

investigations of the interconnection between site locations, environmental features, and their dynamics are limited to small sub-catchments (e.g., Knitter 2014; Pirson 2013).

In this study, several computational analyses are used to investigate hypotheses concerning the strength and control of interaction within the Pergamon Micro-Region to gain insights into a potential interaction network. In particular, we conduct visual analyses, assess the location of sites in terms of geomorphometric characteristics, and evaluate the importance of archaeological sites and their mutual interaction.

Topographic and Archaeological Data Sources

To analyze visibility characteristics and potential routes of traffic and communication, topographic data are indispensable because such analyses depend on digital elevation models (DEMs). Because horizontal and vertical resolution and accuracy have major impacts on results, they need to be stated explicitly. In general, the higher the resolution and smaller the measurement error, the more reliable and precise are the results. In this study we use an open-access DEM to guarantee reproducibility.[3] The elevation data

3 We are not allowed to openly provide the DEM we normally use for analyses due to copyright limitations. It

are provided by the EU Copernicus project as EU-DEM v1.1. It is a digital surface model that fuses SRTM and ASTER GDEM data using a weighted average approach. The EU-DEM v1.1 is an improved version of the EU-DEM v1.0, with enhanced geo-positioning. It has a pixel resolution of 25 × 25 m and a vertical accuracy of 7 m RMSE (EU Copernicus Programme 2016).

To prepare the data for subsequent analyses, raw DEMs were reprojected to WGS84 UTM Zone 35, resampled using a cubic spline method, cropped to the research area, and merged to one file. All such steps were achieved using the gdalwarp function (GDAL Development Team 2017).

Archaeological data used in this study includes a dataset of 19 archaeological sites assembled by the Pergamon Excavation Project (Table 1). The coordinates of the sites can be specified as reasonably precise, as they were all taken in the field using GPS devices.[4] All 19 sites in the Pergamon Micro-Region are farmsteads with towers (*Turmgehöfte*) or larger settlements. There is an ongoing debate whether *Turmgehöfte* should be interpreted as military bases (lookout posts) or central tower houses of larger agricultural estates (for detailed discussion, see Lohmann 2015; Zimmermann, Matthaei, and Ateş 2015). In the case of the site at Denizköy (Fig. 3), architectural remains and location indicate its use as a lookout post (Pirson 2019). Even if some sites show construction characteristic of agricultural buildings only, their height and location in the landscape does not exclude the simultaneous use for signal or communication transmission.

The sites considered in this study were all occupied for different periods of time. The dates of individual sites are derived from survey and excavation reports (Pirson 2007), as well as other publications (Schuchhardt 1912; Williamson 2016). Their chronological accuracy varies because of the heterogeneity of their sources and related research contexts. To investigate interaction networks within the Pergamon Micro-Region, we focus on the Hellenistic period, for which a contemporary occupation of all sites can be assumed.

The general processing of the site dataset was done in R, using the readr package to load the data (Wickham, Hester, and Francois 2017) and the sf package to transform it to a spatial data set (Pebesma 2017).

---

is a digital elevation model from the TanDEM-X satellite mission provided by the German Aerospace Center (Deutsches Zentrum für Luft- und Raumfahrt e.V. [DLR] 2019) and the European Space Agency (ESA) with a resolution of 12 × 12 m per pixel and a vertical accuracy of 2 to 4 meters, depending on the slope (Deutsches Zentrum für Luft- und Raumfahrt e.V. [DLR] 2019).

4 Both DGPS and handheld GPS devices were used, with accuracy ranging from less than 10 cm (DGPS) up to 5 m (handheld GPS using point averaging). However, some of the coordinates were not explicitly collected for this study. For a settlement, for example, the center point may have been measured and not the highest point that would be most suitable for an interaction network. To clarify, further field surveys are necessary.

Visibility Analyses

The study of visibility and intervisibility has long been important in archaeology (e.g., Fraser 1983; Wheatley 1995; Brughmans and Brandes 2017) because of the insights it offers in terms of potential landscape perception (Llobera 2001; Gillings and Wheatley 2001). Because of the visual organization and structuring of archaeological landscapes, visibility characteristics can be regarded as important factors in the location of sites (Turchetto and Salemi 2017). For instance, the choice of location for ancient grave-mounds and their levels of visibility have been central to the reconstruction of social landscapes (Gillings 2009).

The first generation of researchers at Pergamon studied lines of sight between Pergamon and its hinterland (Conze and Schazmann 1911). This practice was later followed by Nohlen and Radt (1978) and Wulf (1999). Their assumptions were adopted in part by Agelidis (2012) and critically examined by Pirson (2008). GIS-based analyses that allowed comparison and verification of results were first conducted by Williamson (2014) and Ludwig (2014).

The method known as viewshed analysis has been applied in many archaeological studies since the emergence of GIS software (Wheatley and Gillings 2002). A simple viewshed analysis calculates the entire field of view of a landscape from one location, or observation point, by assessing whether each grid cell of a digital elevation model is visible or hidden from the observation point (Olaya 2009).

Difficulties of viewshed and intervisibility analyses have to be considered during the calculation and especially during the process of interpreting the results (e.g., Whealtey and Gillings 2000). Probably the most frequently mentioned criticism is the fact that visual analyses are carried out over modern landscape topographies and that paleo-vegetation is ignored (Tilley 1994). Potential ancient vegetation could have major impacts on the result of a visual analysis, apart from seasonal vegetation effects. To mitigate this problem, Ogburn (2006) introduced fuzzy viewsheds that integrate environmental (haze, light conditions, etc.) and object-related (size, contrast, etc.) information in calculations. Instead of binary maps, degrees of visibility are assigned between 1 (clearly visible) and 0 (not visible). Because we lack additional environmental and object information, we use the simple viewshed approach.

Although visibility cannot be equated with control, it can be a precondition of it. For this reason, we calculated single viewsheds and afterward combined them into a cumulative map (Wheatley 1995). Such a map shows the degree to which the landscape can be seen from all sites and thereby represents the area that could be controlled visually.

We calculated the viewsheds using GRASS GIS (GRASS Development Team 2017) and the r.viewshed add-on (Toma et al. 2017). The connection between GRASS GIS and R is provided by the rgrass7 package (Bivand 2017), the raster processing features are provided by the raster package (Hijmans 2017). Because we assume that the sites—or at least parts of them—were used as watchtowers, we set an observer height of 10 m (for tower heights, see Lohman 1993, 152, table 6). In case of additional information regarding individual sites—the types and heights of buildings, for instance—this parameter can be adjusted to gain more realistic results.

We assume that intervisibility between places was the basis for two-way communication and was a constitutive element of a network (Nakoinz 2013). To figure out whether the sites do create a visual network—that acts as a tool for communication due to visual interaction—we extract the visibility information from the single viewsheds into a table of intervisible sites using tools provided in the dplyr package (Wickham et al. 2017). For each pair of intervisible sites, we create a line to allow mapping and easy visual inspection of the potential visibility network using the sf package (Pebesma 2017).

Least-Cost Path Analyses

Roads are important and structuring elements of the landscape. They connect places and often show continuity across periods. Least-cost path analysis offers one possibility to reconstruct road networks. In general, this method is used to identify the most likely routes connecting known locations, implementing assumptions about modes of traffic and ease of travel (Branting 2007; Conolly and Lake 2006; Herzog 2008). In our study it is difficult to reconstruct ancient routes based on today's topography. The reason for this is the highly dynamic geomorphological character of the Kaikos Valley, which caused the main stream and its tributaries to change their courses repeatedly (Schneider et al. 2015, 2017). Furthermore, modern industrial-scale agriculture strongly influences the surface characteristics of the Kaikos Valley and thus the DEM on which our analysis is based. Because of this, least-cost path calculations have to be seen as rough indicators of potential routes between sites, not to be taken as definitive representations of ancient roads. Knowledge of ancient bridges or wayposts, however, has the potential to improve topography-based least-cost path calculations because they help channel potential routes at certain locations.

The least-cost paths in our study were calculated using Dijkstra's algorithm as implemented in the gdistance package (Etten 2017) and are based on Tobler's hiking equation (Tobler 1993), which calculates walking speed on footpaths in hilly terrain in *m/s* (Fig. 4): $W=6e^{-3.5abs(S+0.05)}$, where *W* is the walking speed, *e* is Euler's number, *abs* is the absolute value, and *S* is the slope measured as elevation difference. The function is asymmetric because hikers are assumed to walk fastest on gently downward slopes (*S*=-0.035).

The gdistance approach differs from other least-cost algorithms because it uses conductance rather than costs to find paths (Etten 2017). In our case, these are the fastest paths because we use the hiking function that results in speed values. Based on this approach, it is possible to calculate time-based territories around sites—to map territories

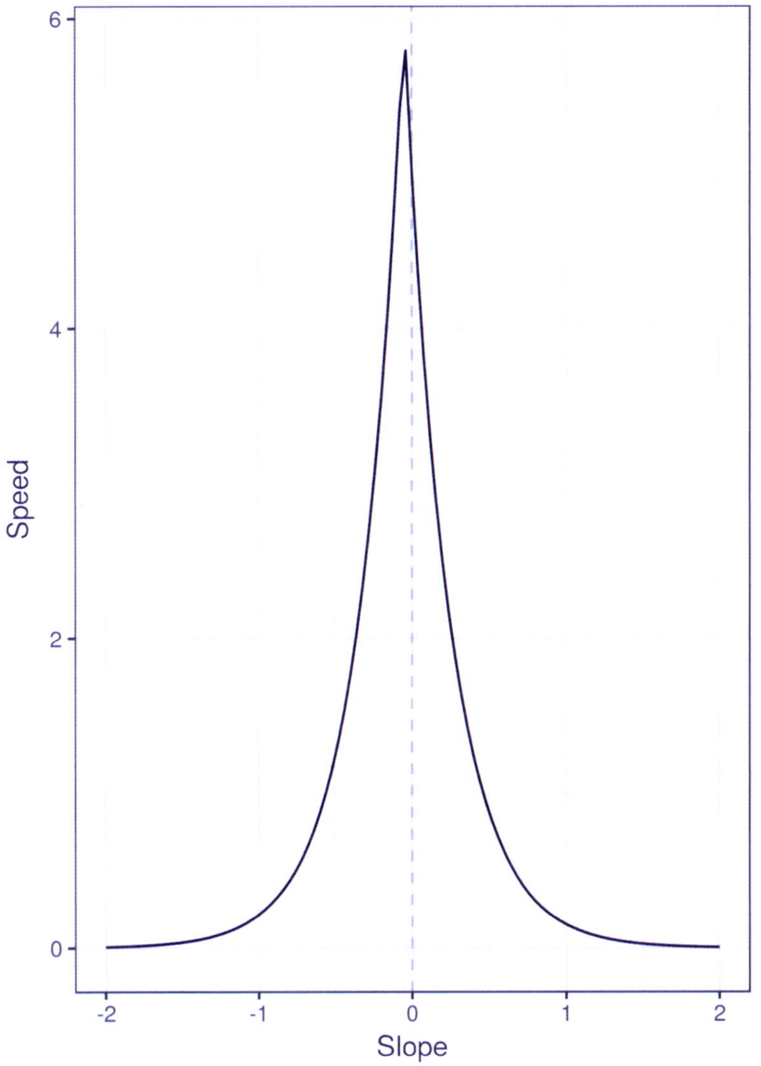

Fig. 4. Tobler's hiking function.

within an hour on foot, for instance. For the purpose of this study, however, we focus only on connections between places and extract their path distances. Calculations were employed between all sites in the dataset, taking each one as both a starting as well as an end point.

The Importance of Sites

Taking the idea of a visual and mobile form of communication further, we investigate the relationship between visual and walking distances. We hypothesize that pairs of sites with shorter visual distance than walking distance are special, because they act as rapid information processing hubs in a network. We use the ratio between the direct (i.e., Euclidean) distance and the walking (i.e., least-cost) distance as the measure of a site's location in *visibility space*. This ratio is then input into a location model of a site's importance. The calculation is based on Wilson's (1971) principle of entropy maximization, applied in archaeological contexts to investigate interaction networks between settlements (e.g., Rihll and Wilson 1987, 1991; Paliou and Bevan 2016; Filet 2017).

We use these ideas to calculate a singly constrained gravity model that acts as a location model (Thomas and Huggett 1980) with the package moin (Knitter 2017). First, the flows (T), that is, the quantity of exchanged goods, people, or interactions, between the different sites i and j are calculated as:

$$T_{ij} = O_i \frac{W_j^{\alpha} c_{ij}^{-\beta}}{\sum_j W_j^{\alpha} c_{ij}^{-\beta}}$$

where O is the size of source location i, and W is the attractiveness of the target location j. Both are set to be equal, because our data offer no information on the size and attractiveness of individual sites. If this information were available, we could additionally weight the attractiveness by a. The distances between i and j are c_{ij}, and these are weighted by β, a standard that in our gravity model equals 2. We understand distance as an abstract concept and refer to it as the ratio between Euclidean and least-cost distances between sites.

After calculating the flows, the importance of a location is given by the amount of flows terminating at the site, that is, (after Thomas and Huggett 1980, 146):

$$D_j = \sum_{i}^{n} T_{ij}$$

Geomorphometric Parameters

The visibility of sites or areas and their connecting paths are highly linked to the surrounding topography. To derive more information on these topographic characteristics, we use two methods from geomorphometry, the science of quantitative land-surface analysis (Pike, Evans, and Hengl 2009, 3).

By investigating the relative topographic locations of sites, we gain insights into the prominence of the different locations. A popular tool to analyze this is the Topographic Position Index (TPI), a measure of the relative difference between the elevation of a focused pixel and the mean elevation of its surrounding pixels (Weiss 2001). Positive values indicate exposed locations, while negative values indicate valley- or depression-like locations (Fig. 5).

Thanks to the possibilities of the raster package (Hijmans 2017), we use a function that calculates the TPI for an area of our choice, in the present case an area with a

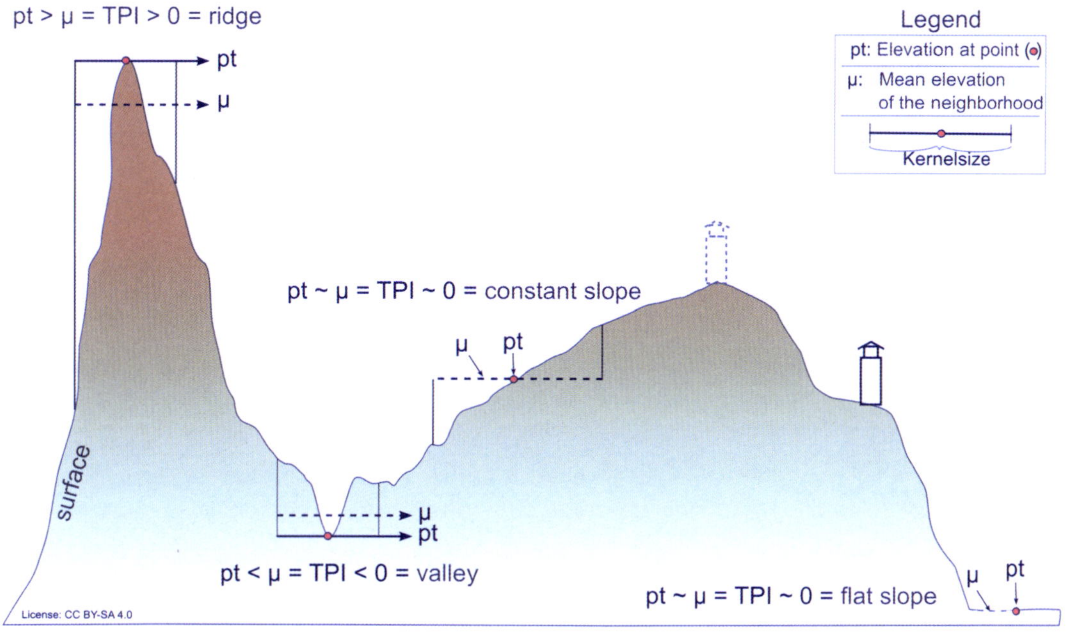

Fig. 5. Principle of the Topographic Position Index (TPI). Positive TPI values are indicative of ridges or hilltops while negative values are representative of valleys. TPI values around 0 correspond to plains or straight slopes. Sketched towers represent archaeological sites: the one with solid lines is a known location, the other is a hitherto unknown site with comparable location characteristics. In this way, TPI can act be a supportive tool in predictive analyses.

diameter of 125 m. We chose this value to represent small, isolated hills. Larger values would represent more extended topographic features and might be useful as well. More detailed field studies would offer empirically based thresholds to improve the analysis.

The other geomorphometric concept we employ is geomorphons (Jasiewicz and Stepinski 2013), which are landform classifications derived from GIS-based topographic classifications that help differentiate topography. Ten different geomorphon-landforms were thus derived from our DEM,[5] using the r.geomorphon algorithm written by Jasiewicz and Stepinski (2013) and implemented in GRASS GIS. We set the search radius of the algorithm to 1 kilometer and the flatness threshold to 1 degree in order to categorize the DEM and gain information on the landforms on which the investigated sites were located. Like in the case of TPI, this was done with no special empirical evidence, hence subsequent analyses recording such information can improve the implementation and its representativeness.

5 The ten were as follows: flat, summit, ridge, shoulder, spur, slope, hollow, footslope, valley, and depression.

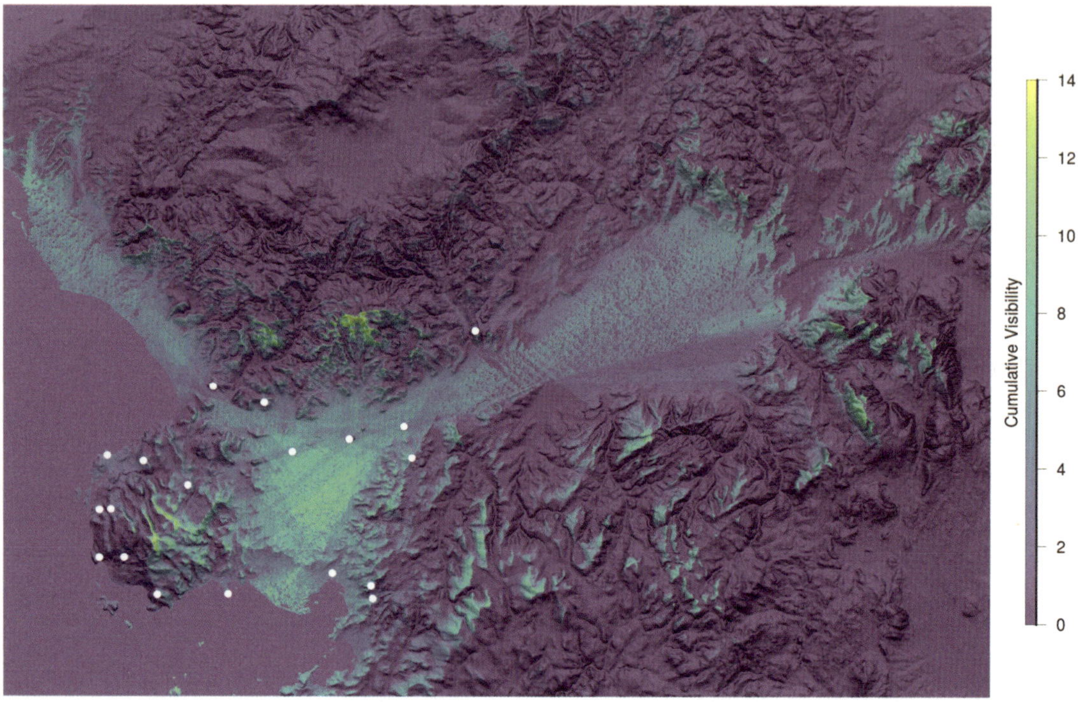

Fig. 6. Cumulative visibility map resulting from viewsheds calculated for 19 sites (white circles). Color ramp created using RColorBrewer package (Neuwirth 2014).

Results

Cumulative viewshed analysis results illustrate levels of visibility for the entire region (Fig. 6). Violet areas with values of zero are not visible from any of the sites. Green and yellow regions represent high levels of visibility. These results imply that the yellow-colored area, the core of the study region, can be viewed from multiple sites. Additionally, the upper slopes and peaks of the Pindasos Mountains in the north are visible to nearly all sites. The same applies to the Aspordenon Mountains in the south and the Kara Dağ Mountains on the Kane Peninsula. The mountains thus form a natural visual boundary or enclosure. Archaeological sites appear to be located around the area of highest visibility in the lower valley.

The results of the intervisibility analysis show a network of mutual visual connections, or a visibility network (Fig. 7). Each line indicates a line-of-sight connection between two sites. A dense network is evident in the plain. There are two main lines on the Kane Peninsula: one runs along the northern edge of Kara Dağ and one along the southern edge. In light of the interpretation of some sites as watchtowers, it is notable that some show no intervisible connections (e.g., the site at Denizköy).

The least-cost paths, or more specifically the slope-determined quickest routes, show two distinct topography-driven distinctions (Figs. 8 and 9; Table 1). In the flat

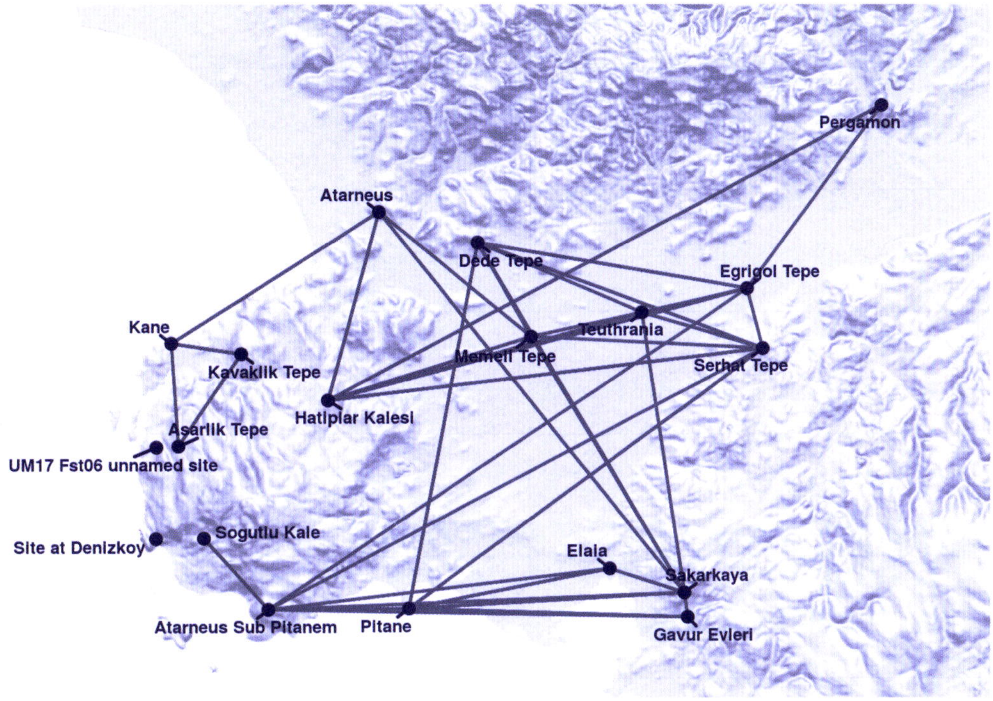

Fig. 7. Intervisibility network. Each line indicates a line-of-sight connection between two sites.

floodplain, where the underlying data contained no obstacles, they are straight lines; in the mountainous region, the paths mostly adhere to valley bottoms or run parallel to slopes.

The analysis of the sites' importance (Fig. 10) shows generally that less accessible sites are more important. This corresponds to the basic intuition that difficult-to-access but easily visible sites are more important in communication networks based on visibility (e.g., Sakarkaya or Teuthrania).

The possibility of interaction depends to a large extent on the location of the sites and thus on the given topography of the micro-region. This is supported by the results of the geomorphometric analyses, where sites are located on summits or ridges (Table 2). All in all, thirteen of nineteen sites lie on summits or ridges. The mean TPI of 2.64 is indicative of elevated locations relative to surroundings.

Discussion

The good visibility of the lower Kaikos Valley from several places suggests that this area could have been visually controlled. This advantageous characteristic is mirrored by the fact that one of the main traffic routes from the south via Elaia to Pergamon, and probably also directly further north, crosses this area (Pirson and Zimmermann 2014). The route was important because it connected Pergamon and its port city Elaia during the Hellenistic period. A large part of Pergamon's trade used this route—to exchange

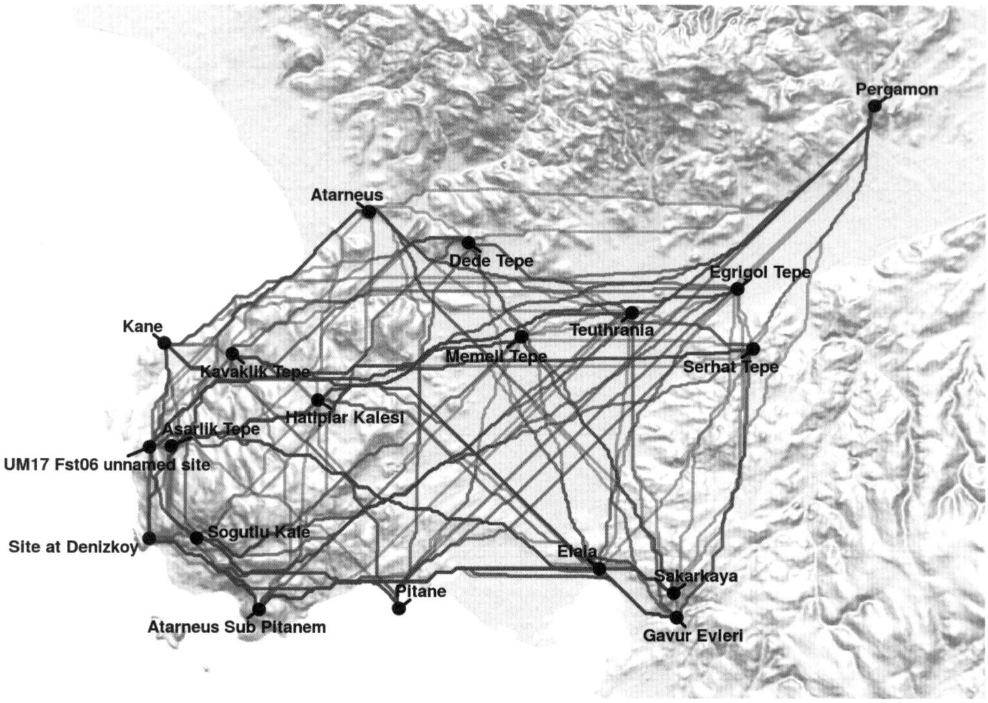

Fig. 8. Least-cost paths between all sites in the Kaikos Valley. The paths are determined by slope and correspond to the quickest paths from each site to any other site.

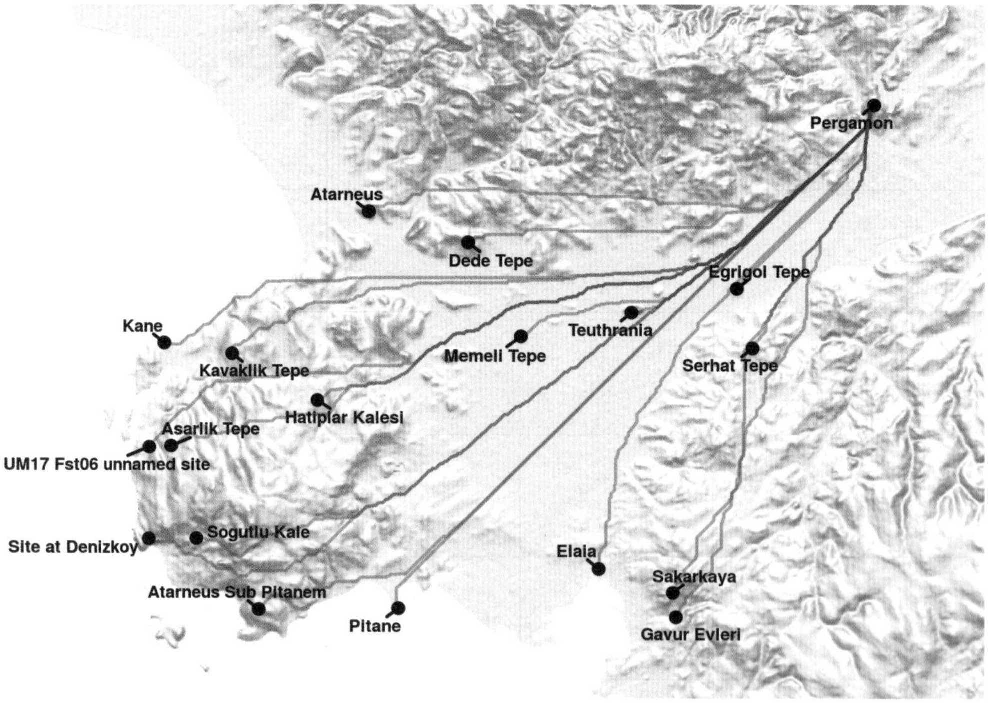

Fig. 9. Least-cost paths terminating at Pergamon. The paths are determined by slope and correspond to the quickest paths from each site to Pergamon.

Table 1. Least-cost path characteristics. Note: Table created using pander (Daróczi and Tsegelskyi 2017).

Name as used in the text	From	To	Least-cost length (km)	Euclidean length (km)	Ratio
Asarlık Tepe	Asarlik_Tepe	Pergamon	38.70	35.16	1.10 1
Atarneus	Atarneus	Pergamon	25.80	23.10	1.12 1
Atarneus Sub Pitanem	Atarneus_Sub_ Pitanem	Pergamon	37.56	35.84	1.05 1
Dede Tepe	Dede_Tepe	Pergamon	21.25	19.20	1.11 1
Eğrigöl Tepe	Egrigol_Tepe	Pergamon	10.89	10.33	1.05 1
Elaia	Elaia	Pergamon	26.18	24.37	1.07 1
Gavur Evleri	Gavur_Evleri	Pergamon	26.92	24.86	1.08 1
Hatiplar Kalesi	Hatiplar_Kalesi	Pergamon	30.87	28.28	1.09 1
Kane	Kane	Pergamon	36.86	33.66	1.10 1
Kavaklık Tepe	Kavaklik_Tepe	Pergamon	34.10	30.91	1.10 1
Memeli Tepe	Memeli_Tepe	Pergamon	20.55	18.97	1.08 1
Pitane	Pergamon	Pitane	31.78	31.22	1.02 1
Sakarkaya	Pergamon	Sakarkaya	25.91	23.89	1.08 1
Serhat Tepe	Pergamon	Serhat_Tepe	13.31	12.30	1.08 1
Site at Denizköy	Pergamon	Site_at_Denizkoy	42.58	38.06	1.12 1
Söğütlü Kale	Pergamon	Sogutlu_Kale	39.68	36.23	1.09 1
Teuthrania	Pergamon	Teuthrania	15.14	14.36	1.05 1
UM17 Fst06 unnamed site	Pergamon	UM17_Fst06_ unnamed_site	39.58	36.06	1.10 1

marble, for example. In addition to the transport factor, this part of the Kaikos Valley played an important role for agriculture and, thus, for the supply of Pergamon (Pirson and Zimmermann 2014). This special importance is also mirrored in the fact that Elaia, among all ports studied, has the shortest path to Pergamon—a decisive factor for the transport of goods between Pergamon and its primary port (Fig. 9 and Table 1).

The absence of a visibility network connecting the whole Kane Peninsula is surprising, yet additional surveys, as in previous campaigns (Pirson 2019), might help close this gap with the discovery of new sites.

The small Hellenistic fortress on Teuthrania (Kalerga Tepe), a well-exposed hill with a high elevation, is generally interpreted as a visual hub for the exchange of information between Pergamon and the coast and vice versa (Williamson 2016, 88). The current study was unable to reproduce this convincing interpretation, although Teuthrania can be seen by one standing on the city hill of Pergamon and vice versa. This inability might result from imprecise coordinates or insufficient DEM resolution, shifting the observation point location to the slope or below the peak, rather than the top of the hill. Another reason might be the representation of each site as a single point rather than areal features.

The results of the geomorphometric analyses shows that the sites are mostly located on ridges and summits and that they have specific TPI ranges. This knowledge could help us search systematically for new, comparable sites in the region in future campaigns: We can now search specific areas with high levels of visibility, corresponding geomorphons, and specific TPI ranges.

On the basis of Elaia and Sakarkaya, a first summarizing conclusion can be made: Elaia, as a trading and naval base, was of great importance to Pergamon in the Hellenistic period (Pirson 2014a). It was therefore important to secure the city and its immediate surroundings. For visual control, the site of Sakarkaya played an important role in this security. It served as a connection into the plain and beyond to Pergamon. At this point it is possible to distinguish two different types of locations within the intervisibility network: (1) locations with a high level of visibility—that is, sites that offer a high degree of visual control based on many intervisible connections and a large field of view; and (2) locations with a low level of visibility—that is, sites that are located in less visible areas with only a few intervisible connections. This differentiation is also demonstrated by their calculated importance and geomorphometric characteristics. Elaia, for instance, is the only locality in a hollow (Table 2) and plays no significant role in the functioning of the intervisibility network (Fig. 10). The city also has no direct line of sight to the plain. Hence, Elaia can be interpreted as one of the locations with a low level of visibility

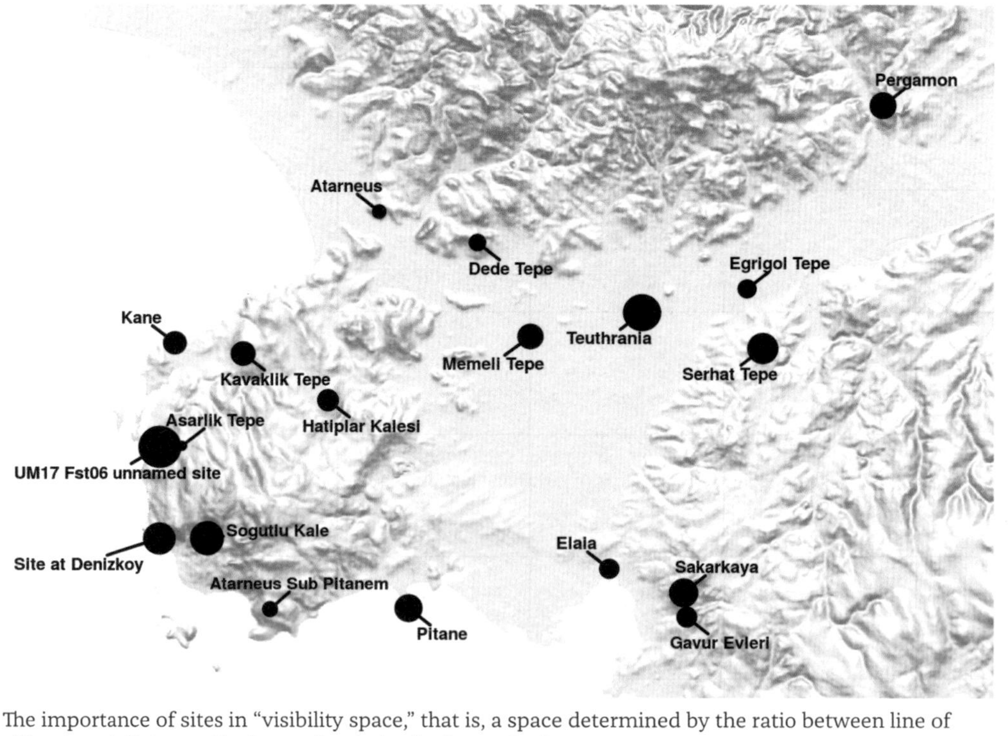

Fig. 10. The importance of sites in "visibility space," that is, a space determined by the ratio between line of sight and least-cost distance. The larger the circle, the larger the importance.

Table 2. Site characteristics. Note: Table created using pander (Daróczi and Tsegelskyi 2017).

Names as used in the text	ID	Name	Easting	Northing	Geomorphon	TPI
Asarlık Tepe	1	Asarlik_Tepe	484250	4316242	ridge	3.65
Atarneus	2	Atarneus	493225	4326889	summit	3.80
Atarneus Sub Pitanem	3	Atarneus_Sub_Pitanem	488251	4308822	summit	2.01
Dede Tepe	4	Dede_Tepe	497650	4325490	summit	4.38
Eğrigöl Tepe	5	Egrigol_Tepe	509698	4323407	summit	4.14
Elaia	6	Elaia	503526	4310681	slope	0.68
Gavur Evleri	7	Gavur_Evleri	506990	4308485	valley	-0.88
Hatiplar Kalesi	8	Hatiplar_Kalesi	490916	4318320	summit	2.59
Kane	9	Kane	483937	4320913	ridge	3.20
Kavaklık Tepe	10	Kavaklik_Tepe	487037	4320431	slope	0.37
Memeli Tepe	11	Memeli_Tepe	500021	4321219	summit	2.76
Pergamon	12	Pergamon	515811	4331730	ridge	3.66
Pitane	13	Pitane	494549	4308876	summit	0.76
Sakarkaya	14	Sakarkaya	506848	4309590	ridge	6.31
Serhat Tepe	15	Serhat_Tepe	510403	4320686	ridge	2.19
Site at Denizköy	16	Site_at_Denizkoy	483245	4312033	slope	-0.37
Söğütlü Kale	17	Sogutlu_Kale	485387	4312050	slope	1.29
Teuthrania	18	Teuthrania	504975	4322311	summit	10.18
UM17 Fst06 unnamed site	19	UM17_Fst06_ unnamed_site	483266	4316192	slope	-0.56

within the network. It is located at the shore, protected in a bay, and can be seen and controlled only from strategic points, such as Sakarkaya. The site of Sakarkaya is located on a ridge (Table 2) and has the second highest TPI of all investigated sites (6.31; Table 2). Its location can therefore be interpreted as very exposed, justifying its important role in the intervisibility network and as a location with a high level of visibility (Fig. 10). It serves as a landmark—for ships, for example—and at the same time as a checkpoint for the city of Elaia and a communication hub to Pergamon.

Conclusions

The aim of this article was to present a landscape archaeological study that follows a strict reproducible workflow and employs solely open data, as well as open-source software. We carried out different landscape archaeological analyses and made their production process accessible by publishing them as a research compendium on a research data repository. Any interested reader can thus assess the reliability of our study and its results. At this point we can conclude that, from a scientific, research-based perspective, the results can be advanced in various ways; further analyses of the Pergamon Micro-Region—including those based on higher resolution elevation data and more comprehensive archaeological information—will offer more far-reaching and interesting insights. Future field surveys will help assess the scientific value of the calculated results and improve their interpretation, especially concerning the analysis of visibility. This study examined only the western part of the lower Kaikos Valley. For a comprehensive analysis of interaction networks within the Pergamon Micro-Region, the eastern part of the lower Kaikos Valley needs to be investigated as well. Until this subsequent work has been done, the present study serves as a valuable indication of the importance of the relationship between Pergamon and Elaia. At the same time, our study shows the advantages of integrating multiple hypotheses. The collaborative examination of different analytical methods and results allows a more comprehensive interpretation. An example of this is the mutual visibility that has already been demonstrated between Pergamon and Teuthrania (Williamson 2016). The result could not be reproduced in this study. This illustrates why it is so important that we share our data and make our results comparable and reproducible. When such computational analyses presented in this article are used in archaeological research, quality criteria, such as the reproducibility of analyses and results, should be considered. If the principles of reproducible research are adopted in the daily routine of archaeological computational analyses, all of us will benefit—today and especially tomorrow.

Acknowledgments

D. Knitter is grateful to the *Collaborative Research Center 1266 - Scales of Transformation* (DFG project number 290391021) for supporting this study.

Works Cited

Agelidis, Soi. 2012. "Die Positionierung von Heiligtümern als Ausdruck der Hierarchisierung von Stadtraum und Landschaft." In *Manifestationen von Macht und Hierarchien in Stadtraum und Landschaft*, edited by Felix Pirson, 99–111. Istanbul: Ege Yayınları.

Bivand, Roger. 2017. Rgrass7: Interface Between GRASS 7 Geographical Information System and R. https://CRAN.R-project.org/package=rgrass7.

Branting, Scott. 2007. "Using an Urban Street Network and a PGIS-T Approach to Analyze Ancient Movement." In *Digital Discovery: Proceedings of the 34th Conference*, edited by Jeffrey T. Clark and Emily M. Hagemeister, 87–96. Budapest: Archaeolingua.

Brughmans, Tom, and Ulrik Brandes. 2017. "Visibility Network Patterns and Methods for Studying Visual Relational Phenomena in Archeology." *Frontiers in Digital Humanities* 4: 1099. https://doi.org/10.3389/fdigh.2017.00017.

Chacon, Scott, and Jason Long. n.d. Git (website). Accessed May 20, 2019. https://git-scm.com/.

Conolly, James, and Mark Lake. 2006. *Geographical Information Systems in Archaeology*. Cambridge: Cambridge University Press.

Conze, Alexander, and Paul Schazmann. 1911. *Mamurt-Kaleh: Ein Tempel der Göttermutter unweit Pergamon*. Jahrbuch des Deutschen Archäologischen Instituts / Ergänzungsheft, vol. 9. Berlin: Reimer.

Copernicus Programme. 2016. "EU-Hydro River Network." Copernicus Land Monitoring Service. https://land.copernicus.eu/pan-european/satellite-derived-products/eu-hydro/eu-hydro-public-beta/eu-hydro-river-network?tab=download.

Daróczi, Gergely, and Roman Tsegelskyi. 2017. Pander: An R "Pandoc" Writer. https://CRAN.R-project.org/package=pander.

Docker. n.d. Docker (website). Accessed May 20, 2019. https://www.docker.com/.

Deutsches Zentrum für Luft- und Raumfahrt e.V. (DLR). 2019. "TanDEM-x Science Server." 23.11.2017 edition. https://tandemx-science.dlr.de/.

Etten, Jacob van. 2017. "R Package gdistance: Distances and Routes on Geographical Grids." *Journal of Statistical Software* 76 (13): 1–21. https://doi.org/10.18637/jss.v076.i13.

EU Copernicus Programme. 2016. "European Digital Elevation Model (EU-DEM), version 1.1." European Environment Agency. https://land.copernicus.eu/pan-european/satellite-derived-products/eu-dem/eu-dem-v1.1.

Faris, Jack, Evelyne Kolker, Alex Szalay, Leon Bradlow, Ewa Deelman, Wu Feng, Judy Qiu, Donna Russell, Elizabeth Stewart, and Eugene Kolker. 2011. "Communication and Data-intensive Science in the Beginning of the 21st Century." *Omics: A Journal of Integrative Biology* 15 (4): 213–15. https://doi.org/10.1089/omi.2011.0008.

figshare. n.d. figshare (website). Accessed May 20, 2019. https://figshare.com/.

Filet, Clara. 2017. "An Attempt to Estimate the Impact of the Spread of Economic Flows on Latenian Urbanization." *Frontiers in Digital Humanities* 3 (10). https://doi.org/10.3389/fdigh.2016.00010.

Fraser, David. 1983. *Land and Society in Neolithic Orkney*. BAR British Series 117. Oxford: BAR.

GDAL Development Team. 2017. "GDAL: Geospatial Data Abstraction Library, version 2.2.3." Open Source Geospatial Foundation. http://www.gdal.org.

Gentleman, Robert, and Duncan Temple Lang. 2007. "Statistical Analyses and Reproducible Research." *Journal of Computational and Graphical Statistics* 16 (1): 1–23. https://doi.org/10.1198/106186007X178663.

Gillings, Mark. 2009. "Visual Affordance, Landscape, and the Megaliths of Alderney." *Oxford Journal of Archaeology* 28 (4): 335–56. https://doi.org/10.1111/j.1468-0092.2009.00332.x.

Gillings, Mark, and David Wheatley. 2001. "Seeing Is Not Believing: Unresolved Issues in Archaeological Visibility Analysis." In *On the Good Use of Geographical Information Systems in Archaeological Landscape Studies*, edited by Bozidar Slapšak, 25–36. Office for Official Publications of the European Communities. http://eprints.soton.ac.uk/28803/.

GRASS Development Team. 2017. "Geographic Resources Analysis Support System (GRASS GIS) Software, version 7.2." Open Source Geospatial Foundation. http://grass.osgeo.org.

Herzog, Irmela. 2008. "Berechnung von optimalen Wegen am Beispiel der Zeitstraße." *Archäologische Informationen* 31: 87–96.

Hijmans, Robert J. 2017. Raster: Geographic Data Analysis and Modeling. https://CRAN.R-project.org/package=raster.

Horden, Peregrine, and Nicholas Purcell. 2000. *The Corrupting Sea: A Study of Mediterranean History*. Oxford: Blackwell.

Jasiewicz, Jarosław, and Tomasz F. Stepinski. 2013. "Geomorphons: A Pattern Recognition Approach to Classification and Mapping of Landforms." *Geomorphology* 182 (0): 147–56. https://doi.org/10.1016/j.geomorph.2012.11.005.

Knitter, Daniel. 2014. "Central Places and the Environment." PhD diss., Freie Universität Berlin. http://www.diss.fu-berlin.de/diss/receive/FUDISS_thesis_000000096525.

———. 2017. "Moin: Modelling Interactions in R." https://doi.org/10.5281/zenodo.841106.

Knitter, Daniel, Hartmut Blum, Barbara Horejs, Oliver Nakoinz, Brigitta Schütt, and Michael Meyer. 2013. "Integrated Centrality Analysis: A Diachronic Comparison of Selected Western Anatolian Locations." *Quaternary International* 312 (October): 45–56. https://doi.org/10.1016/j.quaint.2013.04.020.

Knuth, Donald Ervin. 1984. "Literate Programming." *The Computer Journal* 27 (2): 97–111.

Llobera, Marcos. 2001. "Building Past Landscape Perception With GIS: Understanding Topographic Prominence." *Journal of Archaeological Science* 28 (9): 1005–14. https://doi.org/10.1006/jasc.2001.0720.

Lohmann, Hans. 1993. *Atene: Forschungen zu Siedlungs- und Wirtschaftsstruktur des klassischen Attika = Atēnē*. Köln: Böhlau.

———. 2015. "Wachtturm oder Turmgehöft? Privater oder staatlicher Wehrbau?" In *Turm und Tor*, edited by Barbara Beck-Brandt, Sabine Ladstätter, and Banu Yener-Marksteiner, 249–78. Forschungen in Limyra. Vienna: Österreichisches Archäologische Institut.

Ludwig, Bernhard. 2014. "Kult und Landschaft in Pergamon: Mikro- und Makrolevel in der Analyse und Darstellung von Naturheiligtümern und ihrer räumlichen Einbindung." Master's thesis, University of Applied Sciences Berlin and Freie Universität Berlin.

Marwick, Ben. 2017. "Computational Reproducibility in Archaeological Research: Basic Principles and a Case Study of Their Implementation." *Journal of Archaeological Method and Theory* 24 (2): 424–50. https://doi.org/10.1007/s10816-015-9272-9.

Nakoinz, Oliver. 2013. "Räumliche Interaktionsmodelle." *Praehistorische Zeitschrift* 88 (1–2): 226–57. https://doi.org/10.1515/pz-2013-0008.

Neuwirth, Erich. 2014. RColorBrewer: ColorBrewer Palettes. https://CRAN.R-project.org/package=RColorBrewer.

Nohlen, Klaus, and Wolfgang Radt. 1978. *Kapıkaya: Ein Felsheiligtum bei Pergamon*, vol. 12. Altertümer von Pergamon. Berlin: De Gruyter.

Ogburn, Dennis E. 2006. "Assessing the Level of Visibility of Cultural Objects in Past Landscapes." *Journal of Archaeological Science* 33 (3): 405–13. https://doi.org/10.1016/j.jas.2005.08.005.

Olaya, V. 2009. "Basic Land-Surface Parameters." In *Geomorphometry: Concepts, Software, Applications*, edited by Tomislav Hengl and Hannes I. Reuter, 141–69. Developments in Soil Science 33. Elsevier. https://www.sciencedirect.com/science/article/pii/S0166248108000068#!.

OpenStreetMap. 2018. "Reduced Waterbodies as Polygons." OpenStreetMap Data. http://openstreetmapdata.com/data/water-reduced-polygons.

Paliou, Eleftheria, and Andrew Bevan. 2016. "Evolving Settlement Patterns, Spatial Interaction and the Socio-Political Organisation of Late Prepalatial South-Central Crete." *Journal of Anthropological Archaeology* 42 (June): 184–97. https://doi.org/10.1016/j.jaa.2016.04.006.

Pandoc. n.d. Pandoc (website). Accessed May 20, 2019. https://pandoc.org/.

Pebesma, Edzer. 2017. “Sf: Simple Features for R (version 0.5-6).” GitHub. https://github.com/r-spatial/sf/.

Peng, Roger D. 2011. “Reproducible Research in Computational Science.” *Science* 334 (6060): 1226–27. https://doi.org/10.1126/science.1213847.

Pike, Richard J., Ian S. Evans, and Tomislav Hengl. 2009. “Geomorphometry: A Brief Guide.” In *Geomorphometry: Concepts, Software, Applications*, edited by Tomislav Hengl and Hannes I. Reuter, 3-30. Developments in Soil Science 33. Elsevier. http://www.sciencedirect.com/science/article/B7W58-4THRYD9-9/2/c8edf88078b0cb3168ee80ccbb947911.

Pirson, Felix. 2007. “Pergamon: Bericht über die Arbeiten in der Kampagne 2006.” *Archäologischer Anzeiger / Deutsches Archäologisches Institut, Berlin* 2: 13–69.

———. 2008. “Das Territorium der hellenistischen Residenzstadt Pergamon: Herraschaftlicher Anspruch als raumbezogene Strategie.” In *Räume der Stadt*, edited by Cornelia Jöchner, 27–50. Berlin: Reimer.

———. 2013. “Pergamon: Bericht über die Arbeiten in der Kampagne 2012.” *Archäologischer Anzeiger* 2: 79–164.

———. 2014a. “Elaia, der (maritime) Satellit Pergamons.” In *Häfen und Hafenstädte im* östlichen *Mittelmeerraum von der Antike bis in byzantinische Zeit*, edited by Sabine Ladstätter, Felix Pirson, and Thomas Schmidts, 339–56. Sonderschriften. Istanbul: Ege Yayınları.

———. 2014b. “Wirtschaftliche Ressourcen, ländliche Siedlungen und politische Repräsentation.” In *Pergamon*, edited by Ralf Grüßinger, Volker Kästner, and Andreas Scholl, 58-65. Petersberg: Imhof.

———. 2017. “Die Siedlungsgeschichte Pergamons: Überblick und kritische Revision; Mit einem Appendix von Anneke Keweloh-Kaletta.” In *Istanbuler Mitteilungen*, edited by Felix Pirson and Katja Piesker, 43–130. Istanbuler Mitteilungen 67. Tübingen: Ernst Wasmuth.

———. 2019. “Pergamon: Bericht über die Arbeiten in der Kampagne 2017.” *Archäologischer Anzeiger* 2: 109–92.

Pirson, Felix, and Martin Zimmermann. 2014. “The Hinterland of Pergamon: Economic Resources, Rural Settlements and Political Manifestation.” In *Pergamon*, edited by Felix Pirson and Andreas Scholl, 144–61. Istanbul: Yapı Kredi Yayınları.

R Core Team. 2017. *R: A Language and Environment for Statistical Computing*. Vienna: R Foundation for Statistical Computing. https://www.R-project.org/.

Radt, Wolfgang. 2016. *Pergamon: Geschichte und Bauten einer antiken Metropole*. Darmstadt: wbg Philipp von Zabern.

Rihll, Tracey E., and Alan Geoffrey Wilson. 1987. "Spatial Interaction and Structural Models in Historical Analysis: Some Possibilities and an Example." *Histoire & Mesure* 2 (1): 5–32. https://doi.org/10.3406/hism.1987.1300.

———. "Modelling Settlement Structures in Ancient Greece: New Approaches to the Polis." In *City and Country in the Ancient World*, edited by John Rich and Andrew Wallace-Hadrill, 59–95. London: Routledge.

Schneider, Steffen, Albrecht Matthaei, Marlen Schlöffel, Cornelius Meyer, Mario Kronwald, Anna Pint, and Brigitta Schütt. 2015. "A Geoarchaeological Case Study in the Chora of Pergamon, Western Turkey, to Reconstruct the Late Holocene Landscape Development and Settlement History." *Quaternary International* 367: 62–76. https://doi.org/10.1016/j.quaint.2014.07.020.

Schneider, Steffen, Marlen Schlöffel, Christoph Schwall, Barbara Horejs, and Brigitta Schütt. 2017. "First Stratigraphic Evidence and Absolute Dating of a Bronze Age Settlement in the Bakırçay Valley in Western Turkey." *Journal of Archaeological Science: Reports* 12 (April): 316–22. https://doi.org/10.1016/j.jasrep.2017.02.011.

Schuchhardt, Carl. 1912. "Historische Topographie der Landschaft." In *Stadt und Landschaft*, edited by Alexander Conze, 61–143. Altertümer von Pergamon. Berlin: Reimer.

Schuler, Christof. 2016. "Die Aiolis in regionalgeschichtlicher Perspektive: Zugänge und Schwierigkeiten." In *L'Eolide dans l'ombre de Pergame*, edited by Ivana Savalli Lestrade, 303–19. Topoi orient-occident. Lyon: Maison de l'Orient Mediterranéen.

Sommerey, Kai Michael. 2008. "Die Chora von Pergamon: Studien zu Grenzen, Siedlungsstruktur und Wirtschaft." In *Istanbuler Mitteilungen*, edited by Felix Pirson and Martin Bachmann, 135–70. Istanbuler Mitteilungen 58. Tübingen: Ernst Wasmuth.

Thomas, Reginald William, and Richard J. Huggett. 1980. *Modelling in Geography: A Mathematical Approach*. Totowa, NJ: Barnes & Noble Books.

Tilley, Christopher. 1994. *A Phenomenology of Landscape: Places, Paths and Monuments*. Explorations in Anthropology. Oxford: Berg.

Tobler, Waldo. 1993. *Three Presentations on Geographical Analysis and Modeling: Non-Isotropic Geographic Modeling; Speculations on the Geometry of Geography; and Global Spatial Analysis*. Technical Reports 93-1. https://escholarship.org/uc/item/05r820mz.

Toma, Laura, Yi Zhuang, William Richard, Markus Metz, and GRASS Development Team. 2017. "Addon r.viewshed: Geographic Resources Analysis Support System (GRASS GIS) Software, version 7.2." Open Source Geospatial Foundation. http://grass.osgeo.org.

Turchetto, Jacopo, and Giuseppe Salemi. 2017. "Hide and Seek: Roads, Lookouts and Directional Visibility Cones in Central Anatolia." *Open Archaeology* 3 (1): 69–82. https://doi.org/10.1515/opar-2017-0004.

Weiss, Andrew. 2001. "Topographic Position and Landforms Analysis." *Poster Presentation,* ESRI User Conference, San Diego, California.

Wheatley, David. 1995. "Cumulative Viewshed Analysis: A GIS-Based Method for Investigating Intervisibility, and Its Archaeological Application." In *Archaeology and Geographical Information Systems: A European Perspective*, edited by Gary Lock and Zoran Stancic, 171–85. London: Taylor & Francis.

Wheatley, David, and Mark Gillings. 2000. "Vision, Perception and GIS: Developing Enriched Approaches to the Study of Archaeological Visibility." In *Beyond the Map,* edited by Gary R. Lock, 1–27. NATO Science Series A, Life Sciences. Amsterdam: IOS Press.

———. 2002. *Spatial Technology and Archaeology: The Archaeological Applications of GIS*. London: Taylor & Francis.

Wickham, Hadley, Romain Francois, Lionel Henry, and Kirill Müller. 2017. *Dplyr: A Grammar of Data Manipulation*. https://dplyr.tidyverse.org/.

Wickham, Hadley, Jim Hester, and Romain Francois. 2017. Readr: Read Rectangular Text Data. https://CRAN.R-project.org/package=readr.

Williamson, Christina G. 2014. "Power of Place: Ruler, Landscape and Ritual Space at the Sanctuaries of Labraunda and Mamurt Kale in Asia Minor." In *Locating the Sacred,* edited by Claudia Moser and Cecelia Feldman, 87–110. Havertown: Oxbow/Joukowsky Institute Publication.

———. 2016. "Mountain, Myth, and Territory: Teuthrania as Focal Point in the Landscape of Pergamon." In *Valuing Landscape in Classical Antiquity,* edited by Jeremy McInerney and Ineke Sluiter, 70–99. Leiden: Brill.

Wilson, Alan Geoffrey. 1971. "A Family of Spatial Interaction Models, and Associated Developments." *Environment and Planning A* 3 (1): 1–32. https://doi.org/10.1068/a030001.

Wulf, Ulrike. 1999. "Vom Herrensitz zur Metropole: Zur Stadtentwicklung von Pergamon." In *Stadt und Umland,* edited by Ernst-Ludwig Schwandner and Klaus Rheidt, 33–49. Diskussionen zur archäologischen Bauforschung. Mainz am Rhein: von Zabern.

Zenodo. n.d. Zenodo (website). Accessed May 20, 2019. https://zenodo.org/.

Zimmermann, Martin, Albrecht Matthaei, and Güler Ateş. 2015. "Die Chora von Pergamon: Forschungen im Kaikostal und in der antiken Stadt Atarneus." In *Urbane Strukturen und bürgerliche Identität im Hellenismus*, edited by Albrecht Matthaei and Martin Zimmermann, 193–236. Die hellenistische Polis als Lebensform. Heidelberg: Verlag Antike.

III

Mapping History

The OpenOttoman Initiative and the Challenge of Building an Ottoman Gazetteer

Michael Połczyński

Amy Singer

Introduction

The world of digital, digitized, and online resources is an integral aspect of almost all humanities scholarship in the early twenty-first century. From the simplest catalog search or document composition to the most complex computational analyses, scholars in all fields rely on the capacities and efficiency provided by digital search, storage, and retrieval tools; databases; and computational methods to support their research, publication, and teaching. Digital research tools span a spectrum from increasing the efficiency and precision of long-familiar tasks like note-taking and organizing of metadata, to exploring the potential of newer techniques like distant reading, spatial analysis, and the visualizations of networks and change over time. Digital communications enable cheaper, more varied, and faster interactions among people, facilitating not only research but also review of materials for publications, grants, appointments, and promotions, as well as virtual participation across physical distance and other divides. As in all other fields, disciplines in the humanities take advantage of the sources, publications, reference works, and sharing and teaching platforms that are all increasingly available online. These allow for broader and simultaneous accessibility to all kinds of resources, as well as for lowering the financial cost of access to and participation in teaching and scholarly research altogether.

Yet fields of research within the humanities are not equally digital and digitized. This partly reflects the analog states of their fields at the moments when projects to digitize sources and reference works began. Western European and US history, for example, had many more published collections of documents and reference works than the fields of Middle Eastern and Islamic studies. A similar disparity affects the overall orderliness, cataloging, and accessibility of archives and libraries in the former when compared to the latter. The development of optical character recognition (OCR) capacities is much more advanced for Latin-letter character sets, for example, than it is for Arabic characters, printed or handwritten. This varied pace of OCR development for different character sets has meant that texts in some languages become digitized more quickly and more precisely, making sophisticated searches and analyses practicable and reliable. In other fields, creating a corpus of texts and reference resources, digital and otherwise, is in itself still a challenge and a *sine qua non* to advancing certain kinds of digital research.

Despite a continuous flow of new methods and projects, as well as advances made in OCR, the study of the Ottoman Empire—its history, languages, and cultures—is one large realm of humanities research still at the earlier end of the digital curve. Ottoman studies in general, including its Anatolian spaces, shares much with the study of the Middle East and, with respect to the digital, they are closely aligned. At the same time, Ottoman studies comprises and is an integral component of Mediterranean, Balkan, Southeast European, and Slavic studies, and research on the Black Sea and Caucasus regions. The spatial reach of the Ottoman Empire was vast, and its scholars and sources are dispersed far beyond the borders of the ex-Ottoman lands.

This chapter presents the process by which the project called OpenOttoman initially cohered and developed, and its progress toward becoming a gateway to the digital Ottoman world. OpenOttoman is a transnational digital initiative for scholarship and public history focused on Ottoman studies and committed to being open, inclusive, and accessible. Its public face is openottoman.org. In its broadest framing, OpenOttoman emerges from a critically constructive and generally optimistic attitude to what is now labeled distinctly as "digital humanities." A working definition for digital humanities is usefully formulated as the use "of computation to generate, organize, publish, or interpret humanistic data" (Muhanna 2018, 110). As the methods and resources loosely gathered under this rubric become more familiar and more completely integrated into the toolbox of scholars, it is likely that all scholarship (including research, publication, teaching, and public outreach) will be a seamless melding of digital and analog.

OpenOttoman has two goals. First, it aims to facilitate connections among people, between people and projects, and between people and resources for studying the Ottoman world. The goal of these connections is to enhance individual and collaborative scholarship, to support and expand student learning experiences, and to make the discoveries and insights of scholarship accessible to audiences beyond the boundaries of university walls and academic discourse. A further goal of OpenOttoman is to provide opportunities to define, incubate, and develop the basic tools identified as most lacking for Ottoman studies. The first of these tools is the OpenOttoman Gazetteer, discussed in greater detail below.

OpenOttoman is a web-based portal. It is an ongoing attempt to make sense of digital possibilities and publicize them for the benefit of a disparate and far-flung community of scholars and students, as well as to foster the creation of techniques and datasets specifically, perhaps uniquely, beneficial to the study of the Ottoman world. It was workshopped into existence by a group of scholars who came together in 2015 and 2016 around the following questions: What is digital today in the world of Ottoman studies? Why have certain digital projects in Ottoman studies flourished and succeeded while others faltered and failed? What can be learned from these successes and failures, as well as those in other fields? In what ways can digital capacities be leveraged to create resources that serve the needs of people working in Ottoman studies? What works of reference, databases, and tools for creating, managing, linking, using, visualizing, and analyzing data exist or are most urgently needed for scholars of the Ottoman world? How can a digital environment facilitate cooperative endeavors that cross geographical, political, and disciplinary boundaries? And how can a collective achieve any of this in ways that ensure that the results are sustainably open and accessible?

Ottoman studies is one of the most richly endowed fields in the humanities when considered from the perspective of space, time, and topic. It encompasses a vast geography, a period of some six hundred years when defined most narrowly (c. 1300–1922), and ranges across the study of history, art, architecture, literature, language, philosophy, culture, archaeology, economics, climate, and environment. Recent general histories and reference works emphasize the scope of Ottoman studies (e.g., Finkel 2005; Howard 2017; Ágoston and Masters 2009; Georgeon, Vatin, and Veinstein 2015). In fact, discussions sometimes raise the question, perhaps justifiably, as to whether Ottoman studies can be called "a field" at all. For the moment, however, there still seem to be sufficient reasons to consider the shifting parameters of Ottoman space unified by a single dynastic continuity across time as a loosely unified field of study. As a result, one aim of the OpenOttoman project is to help people and projects around the globe discover each other and connect. One possible outcome of the specific project of the gazetteer may be to consider quite intentionally the question of what made any space "Ottoman" at all.

The known sources for the study of the Ottomans, created both inside and outside Ottoman territories, generally increase in density and diversity with the growth of the state, chronologically and physically. Because of the multiplicity of languages, ecologies, ethnicities, and religions of the people who lived under Ottoman rule, these sources are scattered and heterogeneous (for an introductory overview of sources for Ottoman history, see Faroqhi 1999). Only in the present generation has research on the Ottoman Empire acknowledged more widely the value of Ottoman studies writ large, of research based on the less state-centered and more disparate sources, expanding the parameters of research to include and focus on Ottoman minorities and peripheries. On the one hand, Ottoman studies seems like a field waiting to take advantage of digital capacities now on offer and, on the other, it confronts those same capacities with a host of challenges that make clear where technological limitations still remain.

Imagining a Digital Ottoman Platform

OpenOttoman was first called the Digital Ottoman Platform, a name that had some general aspirations and pointed to an affinity for testing the possibilities of computational technologies for the humanities. The initial inspiration came from a specific challenge, resulting from what might be called "intermediate data." These are the products of deciphering and reading sources, whether literary or documentary, inscribed on paper, ceramic, or stone, containing prose narratives or financial accounts, and transforming their contents into a usable form. In some fields this "usable form" is called data, to distinguish it from the raw information presented in the sources. This intermediate stage is an essential phase of establishing a readable working text from which to continue to an interpretation of the text itself. To some extent the work is mechanical and may one day (in the not-too-distant future) be done partly by a machine that can decipher correctly a reasonable percentage of the original text. The computer will then have become a paleographer. However, even the excellent OCR results of typed Latin-character texts still have flaws today, suggesting that significant imperfections will long be the norm in machine interpretations of handwritten texts in Arabic characters, replete with ligatures.

The great challenge, however, is what happens to these readings once the final product—article, monograph, exhibition, or website—has been completed. How can they be made available as a dataset, even an imperfect one, for wider use? Once the initial research is complete, some or all of the materials used typically languish in unopened drawers or files, unless the scholar who created them continues to incorporate the same materials into other projects or has graduate students interested in using them for additional work. If a scholar's interests shift significantly, then the sources and intermediate materials may never serve them again. And yet they are usable and represent the product of countless hours of skilled work. These data are certainly of interest to anyone curious to explore the grounds for the conclusions proffered in any analytic work; they could also serve a scholar working on related topics or wanting to do a comparative study. At some point, in digitized form, these same texts might be reinvestigated with new methods, whether derived from new directions of research, new computational programs and possibilities, or new machines. The challenge is to discover how this "intermediate data" can be cataloged, curated, and eventually made available openly to students and scholars.

In the past, one practice was to publish the original texts (or a representative sample from among their types) and their transliterations as an appendix to the analytic work. Then, more publishers were amenable to including appendices of sources, and it is still possible to publish a critical edition of an important manuscript. However, such appendices have largely fallen out of fashion, and not every manuscript will attract the scholarly energies necessary to create such an edition. Moreover, it is now perhaps a good decision not to publish such materials in paper editions, as they are far more useful in digital forms, openly accessible online. Even the notion of a critical edition is undergoing a metamorphosis as the possibility of creating a digital edition allows for including all textual variants simultaneously. This obviates the need to make a single choice for the publication, or to privilege one while demoting the others to the notes. One example of such an edition is the Baki Project, which is modeling and testing digital tools and technologies to create an exemplary edition of the Ottoman poet Baki (d. 1600) (Kuru, n.d.).

The most obvious digital solution for making intermediate products available in a usable form is to scan them. Yet several generations of computer hardware and software are likely to roll by in the lifetime of one scholar, such that there is a real risk that scans and data will become inaccessible or unreadable. Examples of projects that continue to exist because they reside on university servers are the early Valley of the Shadow project on the American Civil War (Ayers 2007) and the website of links to resources created by Donald Quataert to accompany his 2006 book, maintained since his death in 2010 (Quataert, n.d.). From the broader perspective of the internet as a whole, one can only hope that the passion and money that drive the work of archive.org will continue to make records of what is and was available on the World Wide Web (Lepore 2015). A website dedicated to a particular project might be useful for demonstrating how the documents were originally deciphered and analyzed, but websites, too, need maintenance of the type that scholars may not have the inclination, time, or resources to provide. The sad fate of many a scholar's research files, digital or analog, after retirement or death, as compared with their publications, is an object lesson for the websites of individuals as well. The value of such papers is often unrecognized by non-specialists who may be the final custodians of such materials, in which case they are discarded (Andras Riedlmayer, pers. comm.). Fortunately, it is increasingly the concern of major research libraries to curate the data produced by their research faculty as a standard part of their activities to maintain their collections. Digital materials are now regularly deposited in the digital repositories of university libraries; it remains to ensure that they are cataloged with sufficiently robust metadata so as to make them easily discoverable in the future.

In contrast, quantitative and experimental scientists in economics, biology, physics, and chemistry have more long-standing practices of making available the methodological protocols and/or the data from their research (see Knitter and Ludwig, this volume). The requirement that they do this is today written into the terms imposed on many grant recipients and journal authors, which is exemplified in the instructions to authors at the *Journal of Economic History*, under "Open Access Policies" (Cambridge University Press 2019). Historians and other humanities scholars, however, do not really think of the notebooks, files, and the other stages of work in progress as datasets. Perhaps we should. The question for Ottomanists is how to create practices to suit the types of sources and methods used in their research.

Although the problem of what to do with intermediate data was what initially motivated one of the authors of this chapter (Singer) to imagine something like a Digital Ottoman Platform, that problem has not yet been addressed by OpenOttoman as a central issue, perhaps because of the complexities involved. Moreover, it turns out that there are many fundamental reference and research tools and resources that are more straightforward to incubate and launch. If created, these would immediately benefit many more people and projects, both in and adjacent to Ottoman studies.

All these challenges are part of the general search for ways in which scholars can take advantage of digital and online technologies to cooperate more regularly and fluidly, and perhaps to include more people who are not professional scholars but who want to participate. Can we create a space and define effective protocols in which to work together and share work? For professional purposes, digital projects may require a cooperative effort among several scholars, in contrast to the more familiar humanities paradigm of one scholar, one article. In such cases, the humanities will need to develop mechanisms for acknowledging the comparable roles of each of several authors, to ensure that each is recognized by the authorities who assess individual scholarship for purposes of funding, hiring, tenure, and promotion.

Initiating the Digital Ottoman Platform

One starting point for concrete discussions about the Digital Ottoman Platform was the 2014 article "Digital Frontiers of Ottoman Studies" authored by Chris Gratien, Michael Połczyński, and Nir Shafir (2014). Gratien is widely known as one of the creators of the *Ottoman History Podcast* (*Ottoman History Podcast* 2014). Michael Połczyński works extensively with GIS in his explorations of the vast space of the Polish-Lithuanian-Ottoman frontier in the sixteenth and seventeenth centuries (Połczyński and Połczyński 2013, 2014). Nir Shafir created the hazine.info website to provide insights and practical information on archives and libraries. Their joint article looked at the field of digital humanities and the digital state of Ottoman studies in 2014 to suggest how Ottomanists could "engage in the digital humanities movement not just as passive consumers of digitized sources and publications but also as active users of digital technologies and new media." They specifically point to the "diverse and disparate topics of study" covered by the study of the Ottoman Empire as well-suited to take advantage of the digitization of sources, digital technologies for exploring and analyzing these sources, and internet-based media that can expand the audiences and modes of academic production and publication (Gratien, Połczyński, and Shafir 2014, 37).

With this aim in mind, the first Digital Ottoman Platform workshop convened in 2015 under the auspices of Professor Sabine Schmidtke, faculty member in Islamic History and the School of Historical Studies at the Institute for Advanced Study in Princeton, NJ. This workshop collected professors and librarians of Ottoman and Islamic studies, geographers, a classical archaeologist, and an engineer (Table 1). Individual presentations of research using a variety of digital methodologies and technologies

Table 1. Participants of the first Digital Ottoman Platform Workshop (8–12 June 2015)

Participant	**Presentation Topic**
Lex Berman	China Historical GIS and China Biographical Database
Jim Newhard	Developing Cross-Regional Syntheses from Regional Survey Data: Opportunities and Challenges
Elias Muhanna	Textual Topographies: Some Notes on Patterned Corpora of Arabic Literary Texts
Nir Shafir	Locating the Early Modern Islamic Archive in the Digital Turn
Will Hanley	Flexible Ontologies for Ottoman Person-Data
Maxim Romanov	Analyzing Arabic Biographical Collections at Scale
Meredith Quinn	E-Books in Seventeenth-Century Istanbul
Eda Özel	Relational Model of Data Collection: Geography of North African Piracy in the Seventeenth Century
Guy Burak	Linking Kemalpaşazade: Ottoman Studies and the Sematic Web
Natalie Rothman	Digitizing Dragomans: Exploring Connectivities, Confluences, and Textual Transformations in an Online Repository
Chris Gratien	Public History and the Middle East
Evyn Kropf	What to Expect from the Crowd and the Surrogate: An Experience with Collaborative Manuscript Description in the Digital Environment
Nina Ergin	Mapping Istanbul's Hamams and Routes of Visual Culture: Two Case-Studies of ArcGIS Use in Research and Teaching
Heghnar Watenpaugh	Digital Heritage in a Time of War: The Destruction of Heritage in Syria and Iraq and the Digital Response
Vahe Tachjian	Reconstructing Ottoman Armenian Life, Culture and Society in the Digital Age: 4 Years with Houshamadyan Project
Grigor Boykov	Ottoman Thrace in H-GIS SEE (South-Eastern Europe)
Jesse Howell	Mobility and Infrastructure in the Ottoman Balkans: Geo-Spatial Analysis and Digital Imagery
Elias Kolovos	Mapping Economic Space in the Ottoman World
Ian Gregory	Geographical Text Analysis: Approaches to Understanding the Geographies in Texts
Nicolas Trepanier	Medieval Peasants in Space: Digital Implications of a Study of Landscape
Michael Połczyński	The Ottoman Inscription Archive: An On-Site Crowdsourced On-Line Digital Geodatabase of Ottoman Dedication Inscriptions
Mark Połczyński	The DOP: Opportunity for a Paradigm Shift?/ Creating a Sustainable Competitive Advantage for the Digital Ottoman Platform

fostered discussions of the advantages and challenges of digital endeavors in Ottoman studies, as well as in the humanities more generally. The presentations focused on the way each person had incorporated digital tools into their research, their successes, and the problems they identified. There was no more precise an agenda than to explore *how* the new digital possibilities could best be leveraged for the benefit of Ottoman studies and the extent to which Ottoman studies has specific needs or constraints that have to be taken into account when designing digital projects.

Each presentation contributed a valuable perspective on what is required to design, initiate, manage, fund, scale up, and sustain a successful project. The workshop emphasized to all its participants that most digital projects thrive as partnerships of skills, enthusiasm, professional standards, and material support. Workshop participants shared their own experiences, frustrations, and critiques, examining carefully past projects and those currently under way. A sample list of the persistent problems that have undermined promising initiatives includes the following: declining or absent human or financial resources; proprietary technologies that become obsolete; datasets that fail to migrate from one format to the next; software that becomes incompatible with the next generation of hardware; and ill-defined or uncertain goals. Successful projects must address specific needs, and it is crucial to understand why any dataset is being compiled. Projects also benefit from critical evaluations at an early stage, based on a pilot project or proof of concept, to uncover any methodological weaknesses before committing extensive resources.[1]

One recurrent theme of the discussions in the first workshop was that basic digital datasets and resources are lacking in Ottoman studies. This led to more operative conclusions. First, digital needs among scholars of the Ottoman world have enough in common such that they can usefully be addressed collectively. Second, a dedicated website is the obvious space in which to coordinate the products of our endeavors, from which to publicize them, and to which to attract interested participants and users. Third, the website can, in the short term, usefully organize a directory of scholars, tagged with their interests and their digital experiences. It can also gather, tag, and curate links to the many existing digital projects, catalogs and other resources specific to Ottoman studies. And it can do the same for general resources in digital scholarship.

Finally, it was decided that the most useful resource of greatest benefit to the largest number of scholars and other people interested in the field would be a gazetteer of attested and geolocated places in the Ottoman world. This will be an ever-expanding resource, a reference work on the places of Ottoman space. It will enable people to identify and corroborate place-names, variants and locations across approximately 5.2 million km^2, six hundred years, twenty-five languages, and eight alphabets. From a practical perspective, the gazetteer can grow in a modular fashion, incorporating regional or source-based contributions as they are submitted. As technical issues are resolved, a more ambitious and complex goal is to incorporate spatial data about

1 Published literature on the digital humanities expands at a furious rate, including journals specific to the field. Following the names of authors, there are excellent recorded presentations to be found online at youtube.com. Published materials that will suffice to lead to more general bibliography include the works of Borgman (2015) and Schreibman, Siemens, and Unsworth (2016), and dedicated journals such as *Digital Humanities Quarterly*, *Frontiers in Digital Humanities*, *Journal of Digital Humanities*, and *Digital Scholarship in the Humanities*. The pros and cons of digital humanities are also regularly argued in the popular press. See, for example, Allington, Brouillette, and Golumbia 2016.

the various units of administrative organization of the Ottoman Empire alongside the collection of placepoints in the gazetteer. As with so many digital projects, the creation of the gazetteer requires (at least) the critical scholarly abilities of trained historians, language experts, geographers, and computer scientists. At the same time, it can incorporate the enthusiastic participation of people with varying degrees of training who can contribute to the basic datasets.

Becoming OpenOttoman

After the collective decisions taken at the first Digital Ottoman Platform workshop, a somewhat different group met in June 2016 for a second workshop, to work more intensively on the website and the gazetteer, and to think in a more focused way about the organizational and financial needs of the project (Table 2). The participants renamed the project OpenOttoman, pointing to its initial inspiration, guiding attitude, and overarching goal.

Table 2. Participants of the Second Digital Ottoman Platform Workshop (20-24 June 2016)

Participant	Presentation Topic
Lex Berman	Multiple Languages, Scripts and Orthographies in Historical Gazetteers
Will Hanley	A Schema for the Ottoman Historical Gazetteer
Kelly O'Neill	If It Isn't Somewhere, It Doesn't Exist: The Challenge of Building a Historical GIS for the Russian Empire
Jeremy Guillette	Technical Tricks for Gathering & Reconciling Data
Nicolas Michel	Historical Geography of Ottoman Egypt
Emily Neumeier	Visualizing an Agro-economic Regime in Ottoman Greece & Albania with GIS
Hakan Karateke	Database for Ottoman Inscriptions: Exploring Digital Possibilities
M. Erdem Kabadayı	Mapping Late Ottoman Population Data in Istanbul and for the Empire
Tom Elliot	The Pleiades Project
Antonis Hadjikyriacou	Spatial History Projects in Crete, Cyprus and Beyond
Boğaç Ergene	Economic and "Large-N" Analysis of Islamic Legal Practice in the Ottoman Empire: A View of the Craft's Kitchen
Serkan Şavk	Istanbul's Early Modern Urban Topography and Mapping its Narratives
Sarah Ketchley	The Emma B. Andrews Diary Project: Recreating a Nile Journey in the Digital Age
Amy Singer	Organizer (Hassenfeld Professor of Islamic Studies, History, Brandeis University, Waltham, MA; Professor Emerita, Middle Eastern and African History, Tel Aviv University)
Sabine Schmidtke	Organizer (Professor of Islamic Intellectual History, Institute for Advanced Study, Princeton, NJ)
Marcia Tucker	Organizer (Librarian, Historical Studies/Social Science Library, Institute for Advanced Study, Princeton, NJ)
Maria Mercedes Tuya	Organizer (DigitalScholarship@IAS/Software Support Specialist, Institute for Advanced Study, Princeton, NJ)

As in the first workshop, the participants were a mixed group in terms of fields and methods. The initial focus was on presentations of new projects, but most time was devoted to the hands-on work of moving the gazetteer and the website beyond their planning phases. Once again, the workshop format created a dynamic of close collaboration and productive interaction, allowing the agenda to shift during the day and from day to day in response to what had evolved immediately before. Discussion included additional resources that could be developed to support and enhance research of the entire Ottoman world, including a directory of monuments, an integrated catalog of published and unpublished manuscripts, or a database of individuals.

We hope that one day the OpenOttoman site will play an important role in reviewing both digital tools and digital products that are relevant for Ottoman history, thereby providing a service to the field and a means to navigate the dense flow of new products. Such reviews would address the benefits and drawbacks of these tools. The site will not serve as a repository itself, however. Individual scholars or research groups will store their data at research library facilities, to date the most reliable long-term solution for preserving, protecting, and curating data. The first experimental repository has been set up at Marquette University, as part of the gazetteer project (Połczyński and Połczyński, n.d.).

The OpenOttoman Gazetteer

The digital gazetteer of Ottoman places was inspired by the absence of a reliable and attested work of reference for the geographies of the Ottoman Empire, including places, place-names, administrative divisions, their variants, and their specific chronologies. The gazetteer team, led by one of the authors of this chapter (Połczyński), is now working on a proof of concept. Its members are learning from and creating partnerships with robust projects like the Pleiades community-built gazetteer and graph of ancient places (Pleiades, n.d.). The goal is to create protocols for contributing data that will make it relatively easy for many people to be involved.

Gazetteers are an old technology. A historical gazetteer, in its simplest form, is a catalog of place-names that can provide corresponding names for the same place over time. They serve as a basic resource for various fields of study. Within the discipline of history, classical studies has long produced printed gazetteers, and the field has proven that digital gazetteers are equally relevant and much more useful. Pleiades and Pelagios illustrate, respectively, a historical gazetteer of the classical world and an "aggregation of projects"; Pelagios uses the locations provided by Pleiades to tag place-names in projects and so link them one to another to be mutually discoverable (see Barker, this volume).

Ottoman studies is lacking not only a digital gazetteer, but critically lacks even a robust printed gazetteer. For an entity that encompassed and affected such vast spaces, this is a huge lacuna. It reflects the extent to which Ottoman studies has developed not as a unified field but as one fractured by its sheer size and shifting boundaries, as well as by political geography and language. The closest publication that attempts to tackle the geographical entirety of the Ottoman Empire is Tahir Sezen's (2006) *Osmanlı Yer Adları*, which contains approximately five thousand places. It is a useful resource for well-known places but does not contain the source references that would allow a researcher to rely on it consistently with any degree of confidence. Otherwise, scholars must rely on local gazetteers, gazetteers such as Pleiades, which cover at least part of Ottoman space in different eras, and other resources, which often lack any linguistic or geopolitical connection to the Ottoman era of that place.

Geographical correlation among Ottoman texts themselves can be challenging even for an experienced researcher already familiar with a specific subject. The variety of disparate spellings, scripts, and individual orthographies makes the identification of unknown proper names a challenge. This challenge multiplies when a toponym comes from a language distinct from Ottoman Turkish or its component Arabic and Persian languages—for example, Armenian, Bulgarian, French, German, Greek, Hebrew, Italian, or Russian. The reconstruction of historical Ottoman spaces, even on a micro level, can be an insurmountable task, particularly if the territory in question has passed into the control of a non-Turkish-speaking modern state. The memory and preservation of Ottoman toponyms, which in any case may never have been in daily use by that place's inhabitants, is weaker the further one gets from the centers of the former empire. This has not stopped geographers and Ottomanists from producing Ottoman gazetteers for their respective regions (Andreev 2013). Where Ottoman toponyms did exist, they have sometimes been actively erased as part of nationalist initiatives to minimize the Ottoman past.

One example of the challenges to realizing an Ottoman gazetteer is the city known to the Byzantines as Adrianople and to the Ottomans as Edirne. An incomplete sample list of variant names for this major Ottoman city offers an introductory glimpse of the challenges of Ottoman toponymy. In antiquity, the city was known as Orestias, Uscudama, or Uskadama. Variant names in Latin characters that refer to the city in the Byzantine, Ottoman, and post-Ottoman periods include: Adrianopel, Adrianople, Adrianopoja, Adrianopol, Adrianopoli, Adrianòpolis, Adrianoupolis, Andrianople, Drinápoly, Drinopolje, Edrene, Endirne, Endrene, Hadrianopolis, Hadrianoupolis, Jedrene, and Odrin. In Hebrew texts the city appears as אדירנה and אדריאנופול ; in Arabic characters, for both Arabic and Ottoman Turkish, as ادرنه, اندرنه, ادرنه بولى ; in Greek as Αδριανούπολη and Εντίρνε; in Cyrillic as Адриануполис, Эдирне, and Одрин; and in Armenian as Ատրանա and Անդրիանուպօլիս.

As this one example shows, an Ottoman gazetteer will have to include alternate names and orthographies in the many languages and several alphabets used to describe places in the Ottoman Empire or non-Ottoman places as they appear in Ottoman texts. Each instance will be referenced and given a chronological scope to the extent possible.

The sources of different spellings and examples of different scripts and orthographies are also provided. At present, the focus of the OpenOttoman Gazetteer project is to develop a user-friendly multilingual interface for contributing and exporting datasets and a robust tool set that will allow users to search the gazetteer for information using a multitude of parameters.

An Ottoman gazetteer can therefore help scholars orient their work geographically with far greater accuracy, refining their understanding not only of Ottoman geography but also of the administrative divisions of this extensive and diverse empire. The digital and open-source nature of the gazetteer will ensure accessibility to this information. Beyond important ethical and economic reasons for ensuring that the gazetteer is open-source, a project of this scope can only succeed by connecting and involving scholars from around the former Ottoman lands and beyond.

Looking ahead to the ongoing development of resources and research projects, the gazetteer will enable them to be linked and discoverable on a much broader scale. For example, Ottoman inscriptions (*kitabe*) are found throughout the ex-Ottoman lands and may mention the name of a monument and its location; yet these inscriptions are mostly unreadable by local residents. The Database for Ottoman Inscriptions is working to collect documentation of these and then create readings and translations of the texts for its website (Database for Ottoman Inscriptions, n.d.). It is an independent project that will also benefit significantly from the gazetteer. Like the OpenOttoman Gazetteer, the Inscriptions database is growing gradually, based on the individual efforts of many contributors from across the geography and chronology of Ottoman history, edited and curated by a team of dedicated scholars.

By linking data in the form of toponyms and location coordinates from a variety of geographically based digital projects, the OpenOttoman Gazetteer would serve as a locus of connectivity for the greater field in the way that the Pleiades gazetteer underpins many digital references, projects, and research publications on the ancient world, now linked through the Pelagios site. Since there is significant spatial overlap between the world covered by Pleiades and that of the Ottoman Empire, cooperating with Pleiades has been an obvious path to pursue for OpenOttoman (Pelagios, n.d.). Equally, Pleiades's geolocators make it possible to discover new connections between Ottoman-era materials and projects and those of the cultures and polities that preceded or followed them in the same geographic spaces.

Several successful and well-advanced projects exist as models for the OpenOttoman Gazetteer to draw on. The first, as mentioned above, is the Pleiades initiative for mapping the Ancient World, "a community built gazetteer and graph of ancient places" (Pleiades, n.d.). The Pleiades project, like most large-scale digital gazetteer projects, started with a robust printed gazetteer. It was based on the *Barrington Atlas of the Greek and Roman World*, originally published in 2000 (Talbert 2000). The *Barrington Atlas* itself drew on several centuries of past scholarship and took two decades and some $4.5 million to produce (Princeton, n.d.). It gave the Pleiades team a critical mass of material that has since been expanded and enhanced by contributors using a variety of additional sources.

Pleiades is a highly respected and stable project based at the Institute for the Study of the Ancient World at New York University. Dr. Tom Elliott, co-managing editor at Pleiades, attended the 2016 OpenOttoman workshop and offered valuable guidance on many of the technical aspects of creating a gazetteer. The geographies of the Pleiades project overlap extensively with those of the Ottoman Empire, making it immediately if indirectly relevant for Ottoman scholars. Our plan in partnership with Pleiades is to make the OpenOttoman Gazetteer data visible and accessible through Pleiades as well as through the OpenOttoman interface. Under the umbrella of Pelagios, an organization supporting linked data in the humanities, projects like Pleiades are now being connected to other gazetteer projects. The Cross-cultural After-Life of Classical Sites (CALCS) project is only one of the initiatives to expand the chronological and cultural scope of Pleiades to include Arabic and Ottoman names (GitHub, n.d.). The OpenOttoman Gazetteer is a natural chronological and spatial companion to these.

The second model on which the OpenOttoman Gazetteer has been able to draw is that developed by Lex Berman of the Center for Geographic Analysis at Harvard University. Berman has already guided a variety of projects, most notably the enormous ChinaGIS and JapanMap (Berman 2019; n.d.). Berman's experience with gazetteers on an imperial scale, with all the attending spatial, temporal, and linguistic challenges, has proven invaluable to the OpenOttoman Gazetteer team in a partnership that began to develop in the first Digital Ottoman Platform workshop in 2015. Here, as with the connection to Pleiades, the OpenOttoman Gazetteer links quite naturally to parts of the East Asian geographies that are the focus of these projects. Collaborating with the developers of preexisting gazetteer projects has fast-tracked the technical development of the OpenOttoman Gazetteer and allowed its developers to focus on issues specific to Ottoman social and spatial contexts.

Pilot

In undertaking a project as large as this, it was absolutely crucial to build a test gazetteer as a proof of concept before embarking on the full project. As Połczyński's area of specialization is the Ottoman Empire's northern frontiers, a small-scale digital gazetteer of Crimea served as the pilot project. The Harvard Imperiia Project, led by Kelly O'Neill, is another geographically and temporally relevant project that could be linked digitally to the OpenOttoman Gazetteer (O'Neill, n.d.). Collaborating with our engineer, Mark Połczyński, we worked from a printed gazetteer, Henryk Jankowski's (2006) *Historical-Etymological Dictionary of Pre-Russian Habitation Names of Crimea*. Jankowski's work contains over one thousand entries for populated places, as well as a map showing the locations of 137 populated places. Of those that have dictionary entries and that are shown on Jankowski's map, 99 have been associated with existing places contained in the GeoNames database (GeoNames, n.d.). These are the populated place entries in our Historical Gazetteer of Crimea (Połczyński and Połczyński, n.d.). Jankowski's map also shows borders for six provinces and forty-three districts, which we have displayed using ESRI's ArcGIS (ESRI, n.d.) (Fig. 1).

The repository of Crimean data we created includes gazetteer data for these populated places, districts, and provinces. Both points (inhabited places) and polygons (districts) have been used to test the viability of including the geometries in the larger project. Additional files show historical routes and paths north into the Pontic Steppe. The gazetteer data exist in three distinct formats: TXT files, KML files that can be displayed using applications like Google Earth, and shapefiles that can be used in ArcGIS and QGIS, an open-access Geographic Information System that allows its users a wide range of functions (QGIS, n.d.). By allowing for the exportation of files in a variety of formats, the OpenOttoman Gazetteer hopes to support research on multiple platforms. This pilot project has shown how a larger Ottoman gazetteer could be created in a modular fashion.

The Historical Gazetteer of Crimea project was undertaken in conjunction with a related project, the digital Gazetteer of Beauplan's Ukraine (Połczyński and Połczyński, n.d.). This digital project is based on primary sources rather than a printed gazetteer. We used the seventeenth-century maps of Ukraine and Crimea produced by French engineer and cartographer Guillaume Le Vasseur de Beauplan who served in the military of Polish-Lithuanian king Wladyslaw IV for nearly two decades (Beauplan 1993). After leaving the Polish Crown's service, Beauplan set himself the task of creating the first

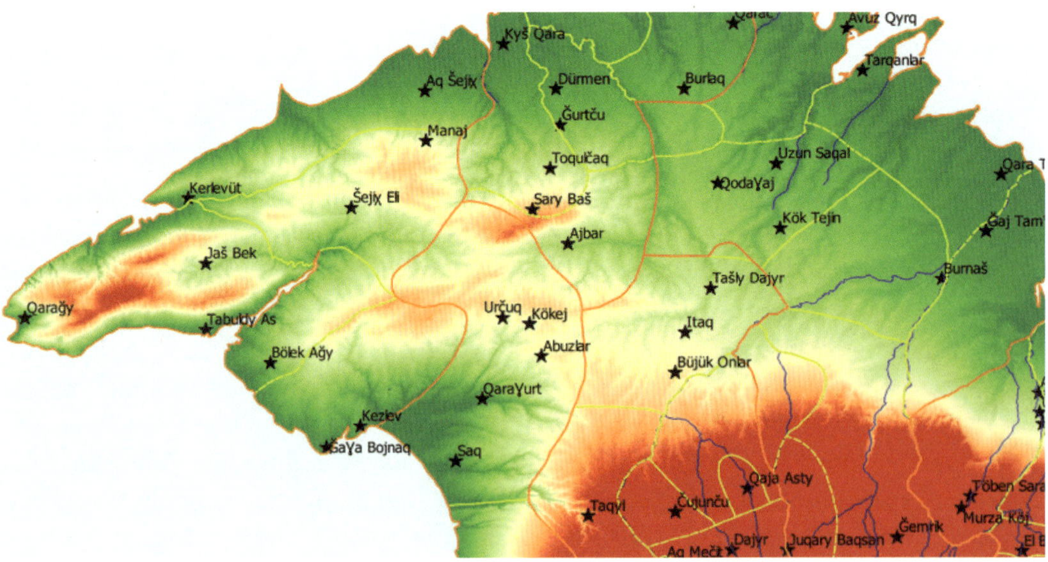

Fig. 1. Screenshot of the Historical Gazetteer of Crimea (HGC) (created by Michael Połczyński).

descriptive maps of Ukraine. Using a selection of these maps, we produced a digital gazetteer that contains 4,158 entries for populated places, rivers, river fords, islands, rapids, forests, mountains, and valleys as shown on nine of Beauplan's maps. This undertaking exemplifies the enormous amount of preparation that is necessary to create usable data from historical documents, for even the most rudimentary historical research. The impetus for creating a digital gazetteer of Beauplan's Ukraine was to support Połczyński's research on the early modern Polish-Lithuanian-Ottoman frontier. In the process of reading documents written in Polish, Ruthenian, Russian, Latin, French, Italian, and Ottoman Turkish, the task of identifying the geographical locations of historical places in the Black Sea region proved difficult. This gazetteer proved indispensable in the study of several subjects related to early modern Ottoman frontier history, including attempts at border demarcation, treaties and frontier violence, and the enormous Black Sea slave trade (Fig. 2).

In approaching these subjects, it was also important to map travel routes in this open steppe frontier zone. The three paths (*szlak* in Polish) in question are the Kuczmański Szlak, Moraski Szlak, and Czarny Szlak. In addition to serving as the main arteries of latitudinal travel in the Pontic Steppe, they enabled the Black Sea

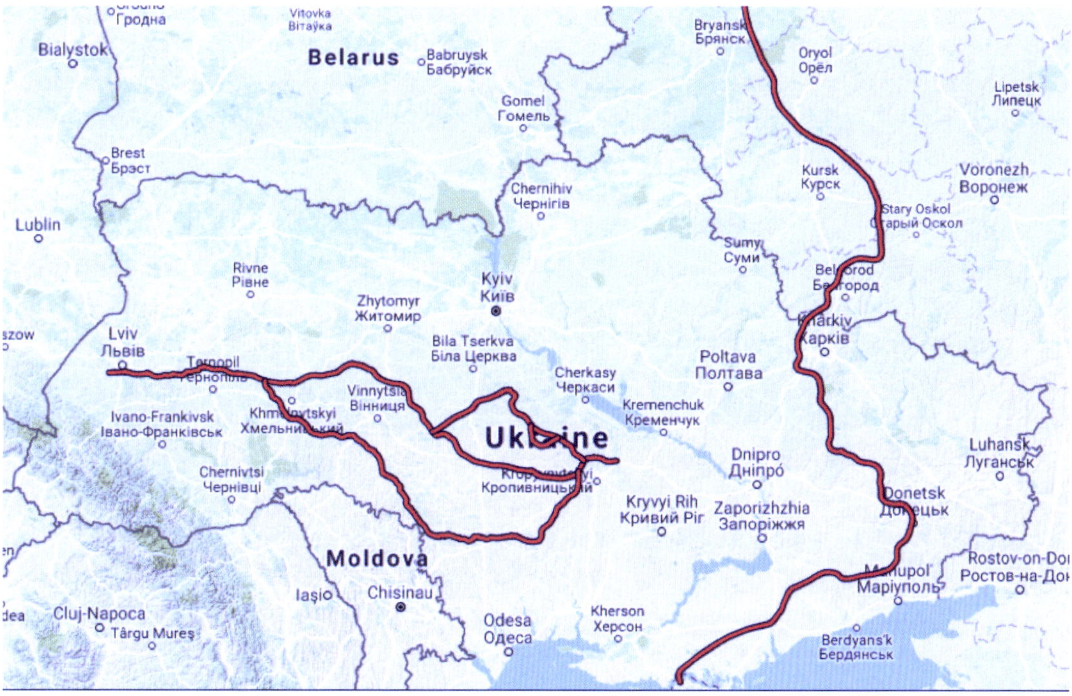

Fig. 2. Looking to the north and using a variety of sources and mapping techniques, the Historical Gazetteer of Crimea and the Gazetteer of Beauplan's Ukraine projects are also reconstructing the historical routes (Kuczmański Szlak, Czarny Szlak, and Moraski Szlak) that fed the Ottoman Black Sea slave trade throughout the early modern period. Success in this has shown us how historical routes in the Ottoman world can be a part of the OpenOttoman Gazetteer.

slave trade in the region (Dziubiński 1996). To trace these paths, several steps were necessary: creating terrain elevation and ruggedness base maps; tracing rivers shown on Beauplan's maps to create a rivers base map; and finding the three thousand five hundred inhabited places appearing on these nine maps in order to create a places base map. The work of producing good base maps had to precede the task of actually tracing Beauplan's travel routes. It took close to three years of research (alongside other routine academic tasks) to produce good base maps. By themselves, these maps are useless. Now that they exist, however, they will serve many different kinds of research projects, including economic history, spatial history, projects on settlement and sovereignty, and virtually any project related to the early modern northern Black Sea littorals that has a geographical or spatial component. Base maps, such as those mentioned here, provide an example of the type and benefits of making materials generally available through OpenOttoman.

Building the Gazetteer: Collaboration

While Ottoman studies is a relatively cohesive field, knowledge of regional sources, in particular non-Ottoman language texts, is typically dispersed among scholars from those regions. To build a critical mass of data most efficiently, we are dividing the empire according to regions of expertise and engaging scholars with the relevant skills and knowledge of sources to help generate robust and comprehensive gazetteer data from the regions they know best. The advantage of this approach is that it can engage scholars and their students in collaborations that reinforce the collegial and international atmosphere of the field. Another advantage is that many of our collaborators are already engaged in the creation of smaller, regionally specific gazetteers for their own research. By implementing a common set of best practices and a uniform entry system, and by developing our ability to execute batch uploading of existing datasets, past and future research of this nature can be incorporated into the OpenOttoman Gazetteer with relatively few problems. With these practices modeled after existing digital gazetteers, OpenOttoman can ensure that its gazetteer will be interoperable with them.

For example, Ottoman places can be identified with place-names from the Classical world. Likewise, as certain territories, like Crimea, changed hands—in this case with imperial Russia—the OpenOttoman Gazetteer can help connect modern spaces with their Ottoman pasts. By linking data from separate projects with shared geolocation coordinates, the gazetteers expand the data available to projects focused ostensibly on common regions separated by politics and time. Moreover, the visualization tools available for spatial data, together with relevant algorithms, make possible sophisticated spatial queries that can help reconstruct Ottoman conceptions of space. Whether looking north and west into its European hinterlands or east and south toward Asia and Africa, we hope that the creation of this tool will help foster stronger interest in interregional connections between the Ottoman world and its neighbors.

In conclusion and fully cognizant of the pitfalls of large collective digital projects, OpenOttoman proceeds with specific interim goals, shaped by its underlying plan and by the people who participate. It does not have a final single target, because possibilities and needs will continue to shift. Through the use of digital resources, it is possible to overcome boundaries of geographical distance, language, and costs that divide scholars who might otherwise work together. Hopefully, such resources will also inspire new kinds of research. We look forward to hearing from people interested in participating in OpenOttoman, bringing skills and ideas, or desiring to link projects to OpenOttoman as it grows. Our address for initial contact is openottoman@gmail.com.

Works Cited

Ágoston, Gábor, and Bruce Masters. 2009. *Encyclopaedia of the Ottoman Empire*. New York: Facts on File.

Allington, Daniel, Sarah Brouillette, and David Golumbia. 2016. "Neoliberal Tools (and Archives): A Political History of Digital Humanities." *Los Angeles Review of Books*, May 1, 2016. https://lareviewofbooks.org/article/neoliberal-tools-archives-political-history-digital-humanities/#!.

Andreev, Stephan. 2013. *Рецнук на Селушну Умена у Названуя на Адмунустратувно-Теруторуалну Едунуцу в Българскуте Зему през XV-Началото на XX Век*. София: Държавна Агенция "Архиви".

Ayers, Edward. 2007. The Valley of the Shadow: Two Communities in the American Civil War. University of Virginia Library. http://valley.lib.virginia.edu/.

Beauplan, Guillaume le Vasseur. 1993. *A Description of Ukraine*. Translated by George Grabowicz. Cambridge: Harvard University Press.

Berman, Lex. n.d. JapanMap. Accessed November 3, 2018. http://worldmap.harvard.edu/maps/397.

———. 2019. China Historical GIS. https://sites.fas.harvard.edu/~chgis/.

Borgman, Christine L. 2015. *Big Data, Little Data, No Data: Scholarship in the Networked World*. Cambridge, MA: The MIT Press.

Cambridge University Press. 2019. "Instructions for Contributors." https://www.cambridge.org/core/journals/journal-of-economic-history/information/instructions-contributors.

Database for Ottoman Inscriptions. n.d. Accessed March 17, 2018. http://info.ottomaninscriptions.com/.

Dziubiński, Andrzej. 1996. "Polsko-Litewskie napady na Tureckie Pogranicze Czarnomorskie w Epoce Dwu Ostatnich Jagiellonów." *Kwartalnik Historyczny* 3: 55.

ESRI. n.d. ArcGIS. Accessed March 17, 2018. http://www.esri.com/arcgis/about-arcgis.

Faroqhi, Suraiya. 1999. *Approaching Ottoman History: An Introduction to the Sources*. Cambridge, UK: Cambridge University Press.

Finkel, Caroline. 2005. *Osman's Dream: The Story of the Ottoman Empire 1300–1923*. London: John Murray.

GeoNames. n.d. Accessed March 11, 2018. www.geonames.org.

Georgeon, François, Nicolas Vatin, and Gilles Veinstein, eds. 2015. *Dictionnaire de l'Empire Ottoman*. Paris: Fayard.

GitHub. n.d. "Cross-cultural After-life of Classical Sites." GitHub. Accessed March 23, 2018. https://github.com/DigiClass/CALCS.

Gratien, Chris. n.d. *Ottoman History Podcast*. Accessed January 30, 2018. http://www.ottomanhistorypodcast.com.

Gratien, Chris, Michael Połczyński, and Nir Shafir. 2014. "Digital Frontiers of Ottoman Studies." *Journal of the Ottoman and Turkish Studies Association* 1, nos. 1–2: 37–51.

Howard, Douglas A. 2017. *A History of the Ottoman Empire*. Cambridge, UK: Cambridge University Press.

Internet Archive. n.d. Internet Archive (website). Accessed May 20, 2019. https://archive.org/.

Jankowski, Henryk. 2006. *A Historical-Etymological Dictionary of Pre-Russian Habitation Names of the Crimea*. Leiden: Brill.

Kuru, Selim. n.d. The Baki Project. Accessed April 5, 2018. http://www.thebakiproject.org/main/.

Lepore, Jill. 2015. "The Cobweb: Can the Internet Be Archived?" *The New Yorker*, January 15, 2015. https://www.newyorker.com/magazine/2015/01/26/cobweb.

Muhanna, Elias. 2018. "What Does 'Born Digital' Mean?" *International Journal of Middle East Studies* 50: 110.

O'Neill, Kelly. n.d. Imperiia: Mapping the Russian Empire. Accessed April 6, 2018. https://worldmap.harvard.edu/maps/russianempire.

Pelagios. n.d. Accessed March 11, 2018. http://commons.pelagios.org.

Pleiades. n.d. Accessed February 7, 2018. https://pleiades.stoa.org.

Połczyński, Michael, and Mark Połczyński. n.d. "Gazetteer of Beauplan's Ukraine." Accessed February 28, 2019. https://epublications.marquette.edu/ottoman_beauplan_gazetteer/.

———. 2013. "A Microsoft Excel Application for Automatically Building Historical Geography GIS Maps." *Transactions in GIS* 17, no. 1: 148–57.

———. 2014. "Using the k-Means Clustering Algorithm to Classify Features for Choropleth Maps." *Cartographica* 49, no. 1 (Spring): 69–75.

———. n.d. "OpenOttoman." Accessed April 25, 2018. http://epublications.marquette.edu/ottoman/.

Princeton University. n.d. "Barrington Atlas of the Greek and Roman World." Accessed March 11, 2018. https://press.princeton.edu/titles/6773.html.

QGIS. n.d. Accessed April 6, 2018. https://www.qgis.org/en/site/.

Quataert, Donald. n.d. Accessed March 17, 2018. http://bingweb.binghamton.edu/~coal/index.html (site discontinued).

———. 2006. *Miners and the State in the Ottoman Empire: The Zonguldak Coalfield, 1822–1920*. New York: Berghahn Books.

Schreibman, Susan, Ray Siemens, and Ray Unsworth, eds. 2016. *A New Companion to Digital Humanities.* West Sussex: Wiley Blackwell.

Sezen, Tahir. 2006. *Osmanlı Yer Adları*. Ankara: Başbakanlık Devlet Arşivleri Genel Müdürlüğü.

Shafir, Nir. n.d. Hazine (website). Accessed February 4, 2018. http://hazine.info/.

Talbert, Richard J. A. 2000. *Barrington Atlas of the Greek and Roman World*. Princeton: Princeton University Press.

A Preliminary Attempt to Construct a Geospatial, Multimodal Ottoman Transport Network for 1899

M. Erdem Kabadayı

Piet Gerrits

Osman Özkan

Turgay Koçak

Introduction

This chapter presents a brief account of work in progress. As a team based at Koç University, we have been working on various aspects of population geography and occupational structure in the Balkans and Anatolia between the Late Ottoman period and the end of the twentieth century within the framework of a broader project (UrbanOccupationsOETR, n.d.). Launched at Koç University in October 2016, our project officially partners and collaborates closely with the University of Cambridge's Cambridge Group for the History of Population and Social Structure (CAMPOP) toward, among other aspects, establishing historical and geospatially accurate multimodal transport networks. Details and some visualizations of preliminary results relevant to historical transport networks can be found on both our and CAMPOP's webpage (CAMPOP, n.d.). The well-known ORBIS project of Stanford University, led by Walter Scheidel and Elijah Meeks and featuring an impressive and interactive webpage (ORBIS, n.d.), was an inspiration both for us and the CAMPOP team.

The expansion and integration of Geographic Information System (GIS) capabilities into academic fields have varied among disciplines and subdisciplines. This uneven integration is evident also for the subfields of larger umbrella groupings, such as the social sciences and humanities. For obvious reasons geography and cartography have been at the forefronts of the geospatial, or GIS, revolution, with their closer ties to geomatics. To be sure, revolution is a strong term in social sciences, and time will tell whether there will actually be a revolution. Such a revolution was argued enthusiastically already in the early 2000s. As Dobson rightfully assessed in 2004, however, GIS and its effects were not revolutionizing the science, and the process could be seen more as a reluctant evolution (Dobson 2004, 573). As historians, we prefer with conventional comfort to wait, see, and comment on change that has already happened. Nevertheless, from our point of view, it is beyond doubt that we are at the verge of a more decisive spatial turn in the field of history, with increasing numbers of examples of the incorporation of geospatial capabilities in historical methods and with ever-growing digital toolboxes. We think the Coursera online course of Pennsylvania State University on "Maps and Geospatial Revolution" in the subfield of Data Science and Data Analysis is a good example (Coursera, n.d.). For a concise evaluation of cooperation between historical studies and GIS in the early 2010s, see Martí-Henneberg (2011, 1–8). Yet, we argue that this spatial turn has not yet been georeferenced, at least not in the field of history. To georeference this new spatial turn, space, in general, and locations and territories, in particular, should be incorporated into historical analyses as geospatially tagged, coordinated places.

We are convinced that this crucial, yet mammoth, task can be achieved only with a digitally well-connected, healthy, and strong development of the emerging interdisciplinary subfield of Historical Geographic Information Systems (HGIS) (Weiwei and Logan 2017), which is being used ever more frequently in historical research (Gregory, DeBats, and Lafreniere 2018). Although more than ten years have passed since the first major publication on HGIS (Gregory and Ell 2007), the fact that no international HGIS journal has been developed shows how HGIS is still in its early stages of development as a research field. This chapter is the result of our modest attempt to contribute to the development of HGIS applications within Ottoman studies, and ours is not the first such effort. Since 2014 a few studies have begun to position Ottoman studies within the emerging field of digital and spatial humanities. In addition to Połczyński and Singer (this volume), we can mention the work of Gratien, Połczyński, and Shafir (2014) and Hadjikyriacou and Kolovos (2015). Experts in HGIS have also already contributed to literature with highly developed analyses of historical transport networks focusing on Balkan railways after the 1870s. Stanev's publications are pioneering in this regard, for example (Stanev 2013). Although preliminary, we think our attempt to construct a multimodal transport network for the Ottoman Empire is timely and, to our knowledge, unprecedented.

Multimodal Components and Their Importance

In constructing digital, georeferenced historical transport networks, multimodality is a crucial element. Without allowing for goods and people to switch modes of transport (coastal navigation, land roads, railways, etc.), advanced techniques of GIS path determination can result in misleading routes. The exclusion of sea travel, in particular, might result in extremely problematic projections of historical transport activity. ORBIS showcases the importance of multimodality very successfully by incorporating fourteen different modes of road travel, in addition to fluvial, coastal, and overseas travel. ORBIS also provides clear information on the functioning of their transport network model in minute detail (ORBIS, n.d.). The preliminary model we present here is drastically primitive compared to that of ORBIS yet includes the crucial characteristic of multimodality even in a very simple form. The four modes we include are two types of roads (primary and secondary), railways, and coastal navigation. The model is based on two historical maps from 1899, a lighthouse directory from 1911, and a commercial vessel logbook from 1894. In the following sections we explain, step by step, how and based on which material we have constructed the 1899 historical transport network. First, we illustrate the geographical coverage of the exercise.

Scope and Geographic Coverage of the Network

Although on land our case study is limited to the 1899 borders of the Ottoman Empire, on sea and regarding the coastal maritime network, we go beyond Ottoman borders. Our scope encompasses almost the entire coastline in Figure 1, except for certain parts of today's Greek coast. In constructing the network, we used two historical maps as the main building blocks. These maps were made in the same year, 1899, and at the same scale (1:1,500,000). For the sake of simplicity, we call them the Huber (Huber 1899) and Vogel maps, the latter derived from the 1908 edition of the map created with the help of B. Domann in 1899 (Vogel and Domann 1908). Both maps were georeferenced by and downloaded as GeoTIFF files from the EarthWorks geospatial data discovery application (EarthWorks, n.d.) of Stanford University Libraries, where details of their georeferencing can be found. We made only minor adjustments regarding the datum and projection of the maps to bring them into alignment with preexisting project data (Fig. 2).

Coastal Navigation Mode

Turning to the modes of transport incorporated in the network, we began with coastal maritime transport. First, we "map mined" nearly 500 harbors, 246 from the Huber and 245 from the Vogel maps. In doing so we aimed to geolocate all possible nodes of maritime transport. From both maps, we included all points on coasts that are connected to inland roads (Fig. 3).

Thereafter, we aimed to construct a corridor of safe coastal navigation. We based the outer boundary of that corridor on lighthouse visibility ranges. To do this, we extracted latitude and longitude data from a 1911 directory of lighthouses (Fener Risâlesi 1911). This printed source in Ottoman script is a partial translation of a non-specified contemporary European directory of lighthouses for the entire Mediterranean, and it includes 179 lighthouses for our chosen territory (Fig. 4). The lighthouse directory also provides detailed information about the visibility range of individual lighthouses (Fig. 5), which, when combined, served to set the outer boundary of safe coastal navigation (Fig. 6).

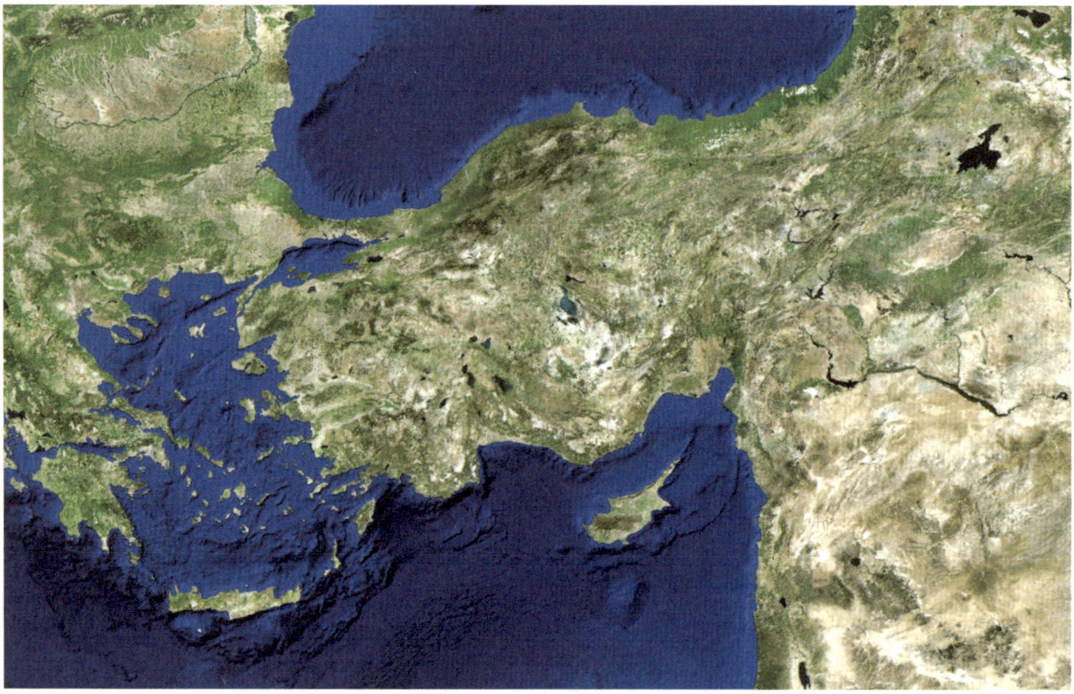

Fig. 1. Scope of the network.

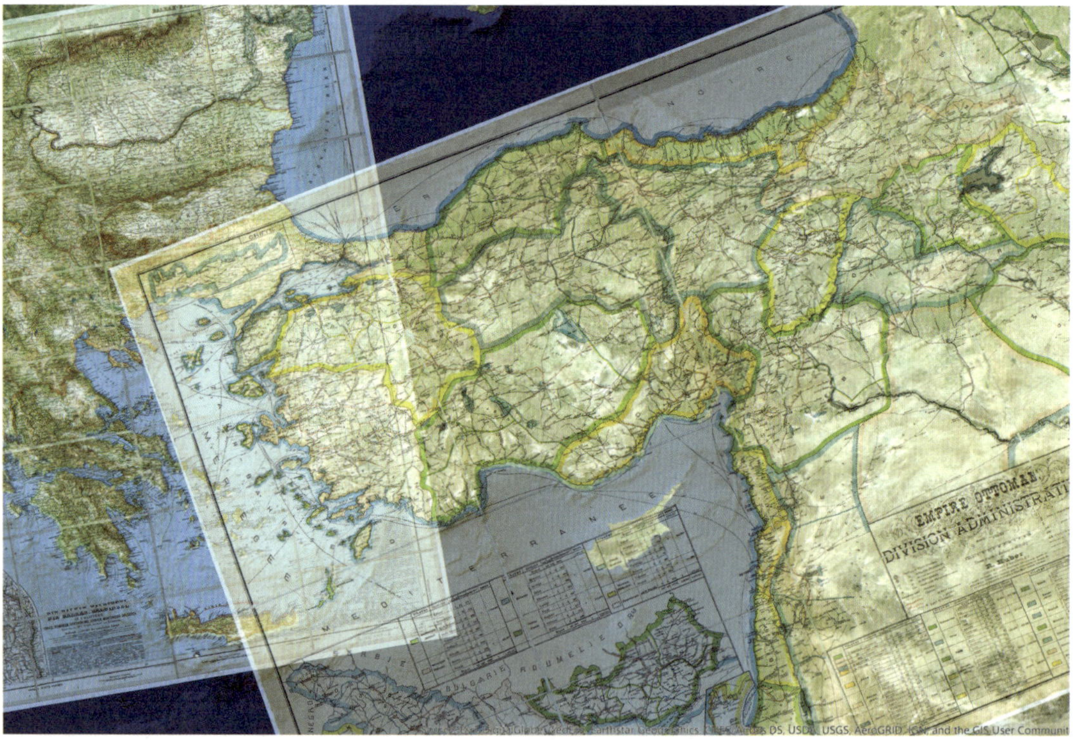

Fig. 2. Georeferenced Huber and Vogel maps.

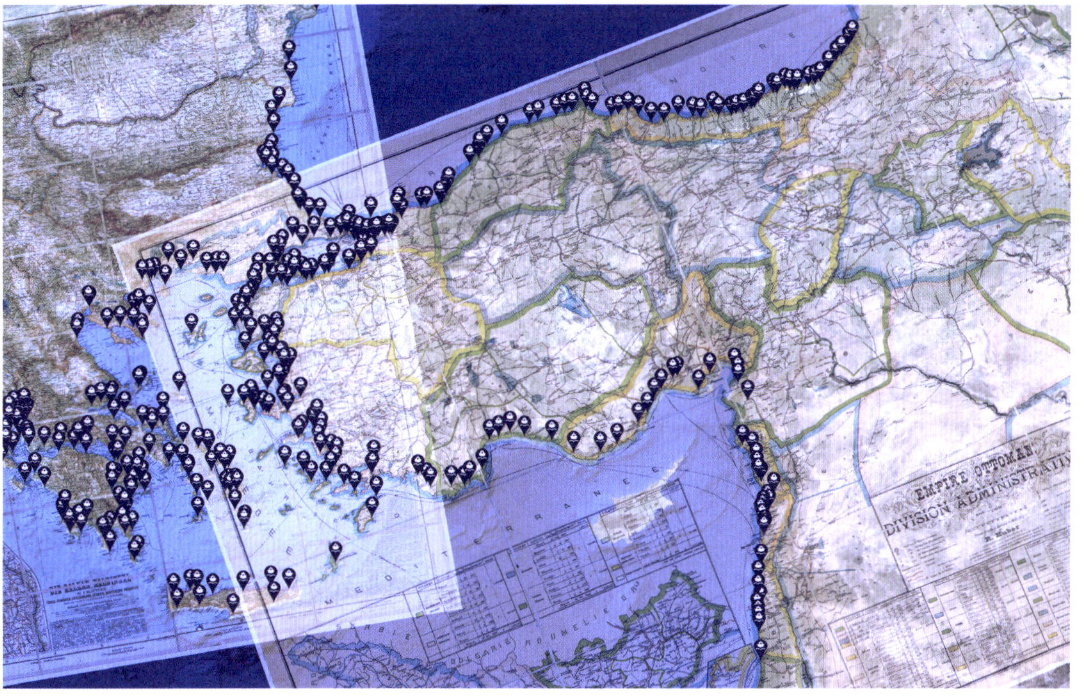

Fig. 3. Possible nodes of maritime transport.

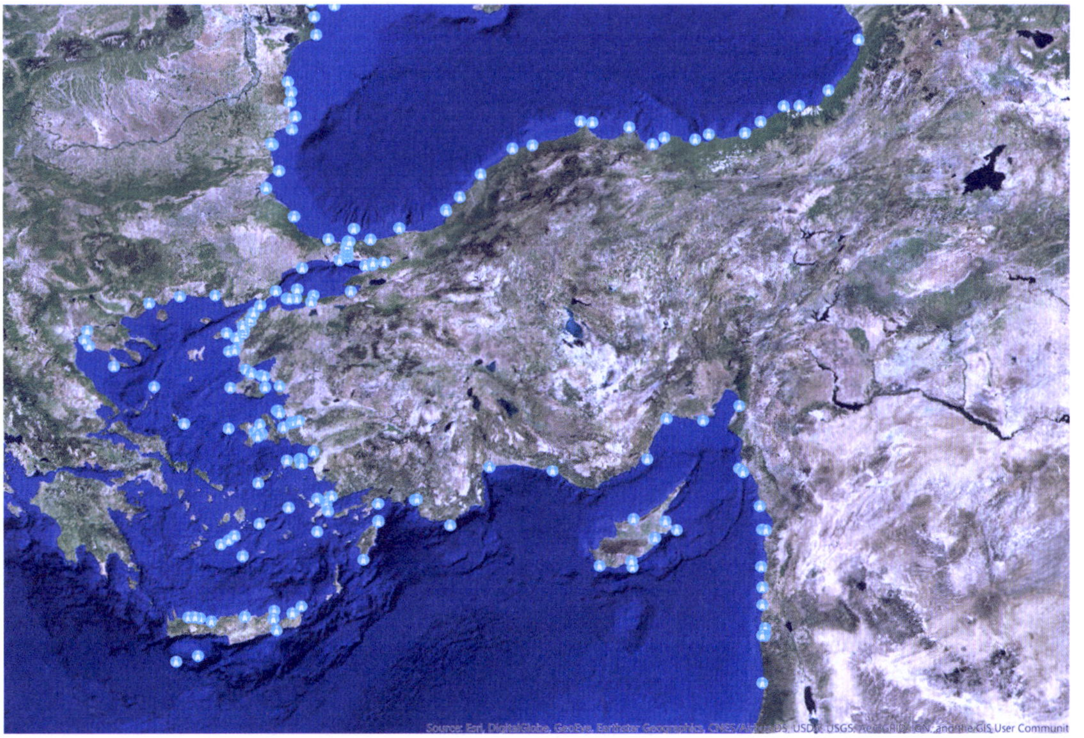

Fig. 4. One hundred seventy-nine lighthouses from *Fener Risalesi* (1911).

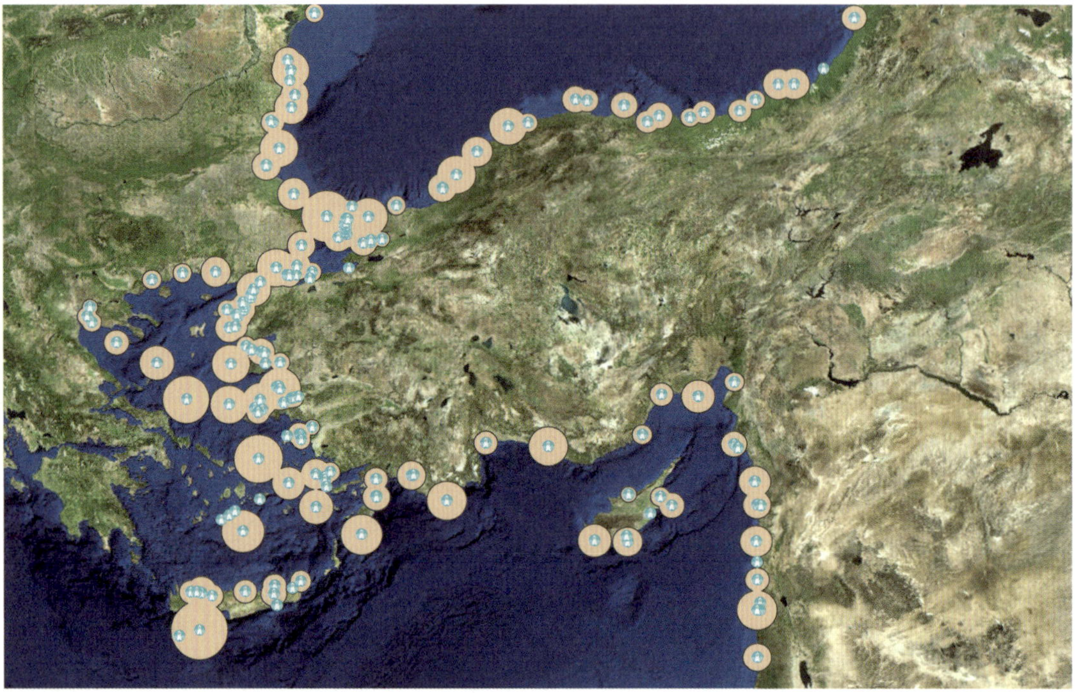

Fig. 5. Visibility ranges of 179 lighthouses from *Fener Risalesi* (1911).

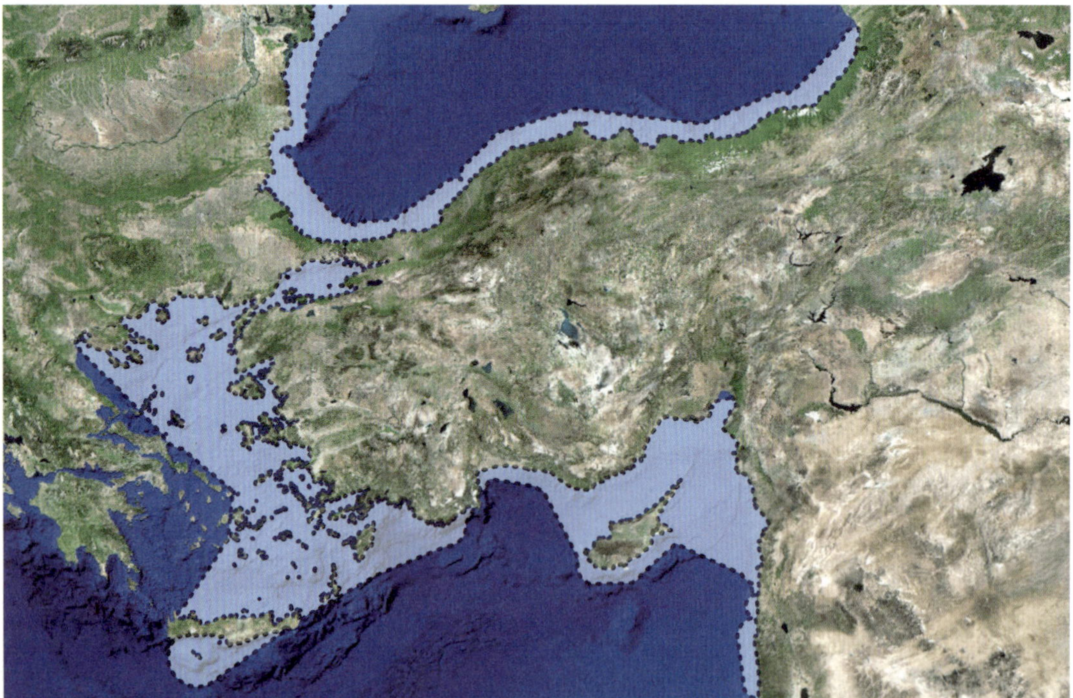

Fig. 6. Coastal navigation corridor.

For the inner boundary of the safe coastal navigation corridor, we made combined use of bathymetric data from the European Marine Observation and Data Network (EMODnet, n.d.). For our analysis we resampled the EMODnet data to present a higher spatial resolution of 30 meters, and we set 8 m as the minimum depth for safe navigation (the minimum depth for passage and embarkment of vessels) based upon historical documentation from the 1900 restructuring and modernization of the Salonica harbor (HH.THR. 214/30). This document from the Ottoman state archives is quoted in Bayram (2009, 158). We are aware of the fact that in several Ottoman maritime transport nodes barges were used to load and unload vessels. Nevertheless, we think a minimum depth of 8 m still makes sense in this early stage of our model development. The spatial resolution of EMODnet bathymetry data is also unfortunately not perfectly in accord with our needs, yet for this preliminary version, we think it serves our purposes. Combining these data, we created the safe coastal navigation corridor, displaying impassable parts in red (Fig. 7).

The Istanbul Naval Museum Archives belonging to the Maritime History Archives of the Turkish Naval Forces (MHATNF) has a rich collection of detailed vessel logbooks. To test our coastal navigation corridor, we used the logbook of a steamer named the *Turk* (MHATNF, 4118). In particular, we closely examined three journeys of the *Turk* from Istanbul in 1894, one each to Rize (following the southern Black Sea coast), Salonica, and Jaffa (Fig. 7).

The logbook of the *Turk* provides exact information on the mileage between stops and the times of arrival to and departure from stops in these three representative journeys. We compared the mileages given in the logbook with the distances our model provides, constraining travel within the safely navigable zone of the model. In general, time distances given in the logbook deviated no more than 10 percent between stops in these comparisons. The best fit was the longest journey, between Jaffa and Istanbul, with fifteen stops (Fig. 8).

Land Transport Modes

After establishing a workable coastal navigation corridor, we extracted land transport features as strings from the Huber map, that is, roads and railway tracks, in three categories (with the French terms deriving from the map legend): primary roads (*chaussées*), secondary roads (*chemins*), and railroads (*chemins de fer en exploitation*). *Chaussées* were paved roads constructed by laying and compacting successive layers of broken stone. An authoritative and contemporary dictionary gives the meaning of *chaussée* as a macadamized road, after John Loudon McAdam, who revolutionized the construction of paved roads in England in the early nineteenth century (Redhouse 1890, 1141). Based on this information, and more importantly via analyzing the locations and density of the *chaussée* network, we ranked *chaussées* as primary roads. *Chemin*, on the other hand, was translated into Ottoman as path or road without any further

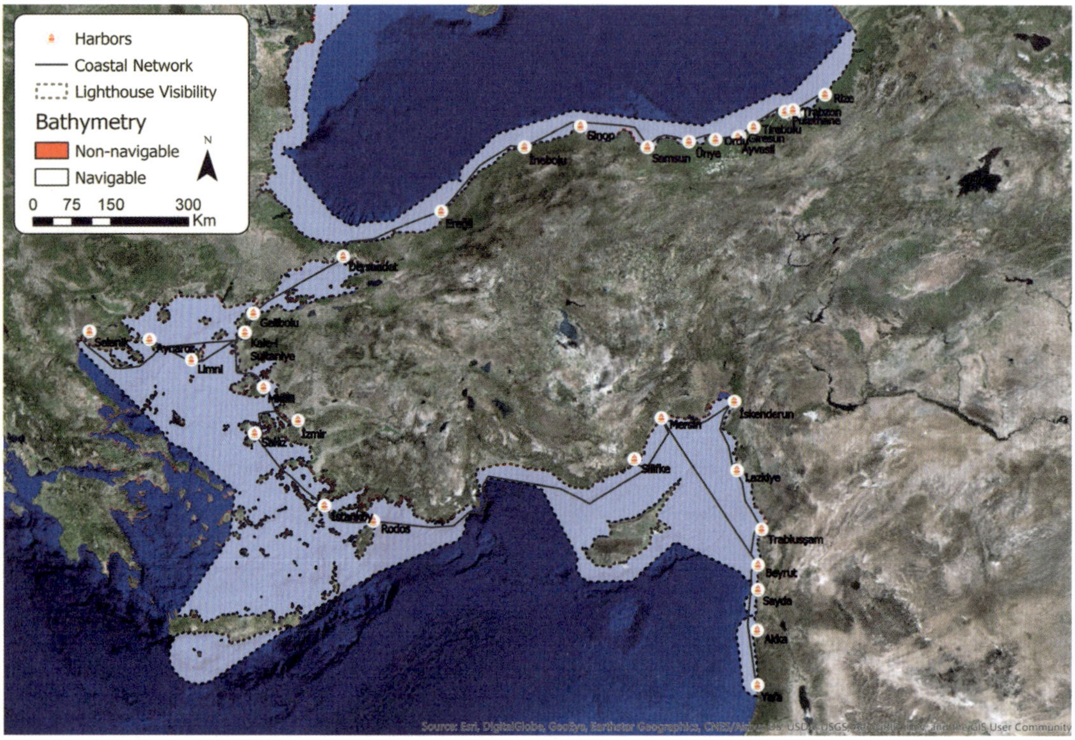

Fig. 7. Return journeys (from/to Istanbul from/to Jaffa, Salonica, and Rize) of the steamer *Turk*, 1894.

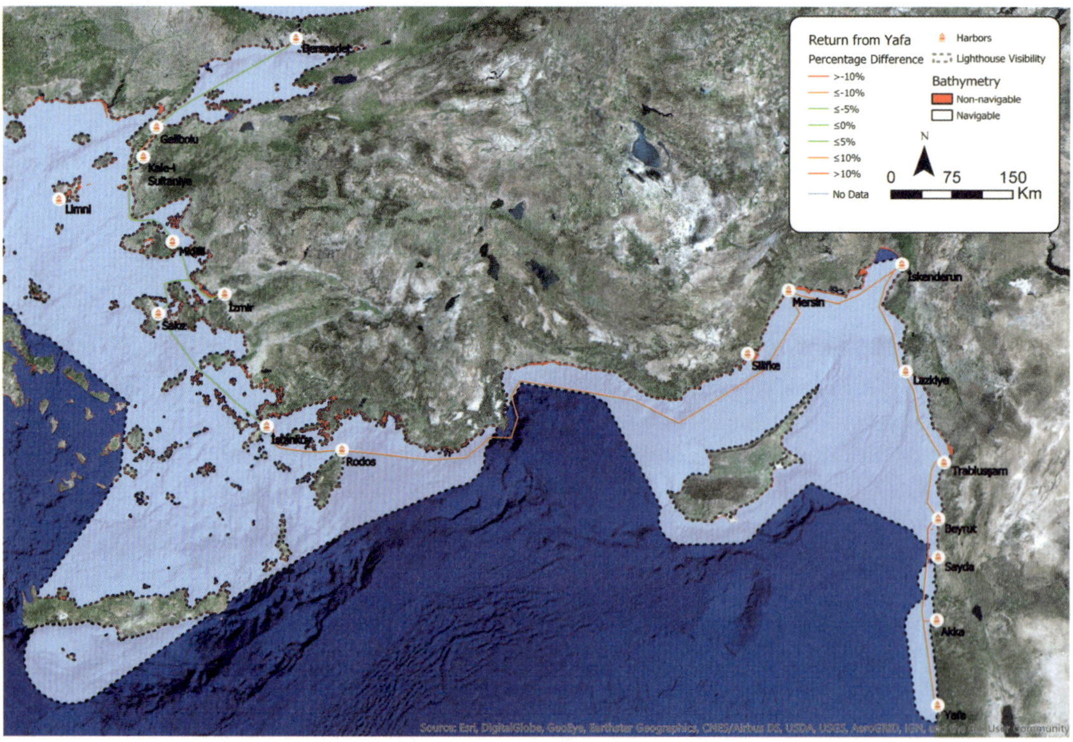

Fig. 8. Close-up of Jaffa–Istanbul journey of the steamer *Turk*, 1894.

qualification in the most used contemporary French-Ottoman dictionary (Sami 1901). Therefore, we categorized *chemins* as unpaved, conventional roads of secondary connections, and grouped them into secondary roads. The third and the last category of polylines extracted from the Huber map were railroads in operation. During the digitization of strings/polylines of these three categories from the Huber map, we also digitized every point of their intersection, as nodes, to allow multimodality and movement among primary and secondary roads and railroads. Due to relatively limited coverage of railroads and lacking detailed location information for railway stations, we included railroad intersections with primary and secondary roads as possible locations to switch between modes. In total we connected around two thousand nodes via these three modes of land transport (Fig. 9). Combining this land transportation data with the coastal navigation corridor allowed us to construct the first version of the 1899 multimodal transport network model for the Ottoman Empire (Fig. 10).

Travel Times, Distances, and Average Speed of Modes

As a last step in this exercise, we aimed to calculate the average speeds of each of the modes of the network. Because the network is georeferenced, we can calculate distances in kilometers using the road connections between the nodes of the model. The Huber map also provides travel times in hours for primary and secondary roads. Therefore, we were able to calculate average speeds on primary and secondary land roads solely relying on the Huber map and the model. We have time and distance data for 327 primary and 899 secondary road segments of the Huber map. We extracted respective road segment lengths from our model and calculated average speeds for both road types as 4.8 kmph for primary roads and 4.7 kmph for secondary roads. As a result, we rounded up the general average speed to 5 kmph for both road types. For railroad transport we used an 1875 timetable of the Ottoman Railway Company (n.d.) for the Izmir-Aydın segment to calculate the average speed of trains as 22 kmph. For the average speed of coastal navigation, we relied on the 1894 logbook of the *Turk*, which traveled at an average rate of 17.5 kmph over 54 segments. With average speeds for all modes of transport calculated, we are able to calculate average travel times between any connection among the nodes of the network based on distances and average speeds. For example, the model allows us to simulate a hypothetical journey through the network from Eskişehir to Baghdad using all possible modes of transport. The resulting simulated journey would stretch 2,028 km over 15 stops, would take 12 days and 9 hours of active travel time (ignoring the passage of time while switching modes), and takes advantage of both overland and coastal routes (Fig. 11).

In Lieu of a Conclusion, Future Plans

With this brief chapter we wanted to share the very preliminary results of our efforts to construct a geospatial, multimodal Ottoman transport network for the late nineteenth

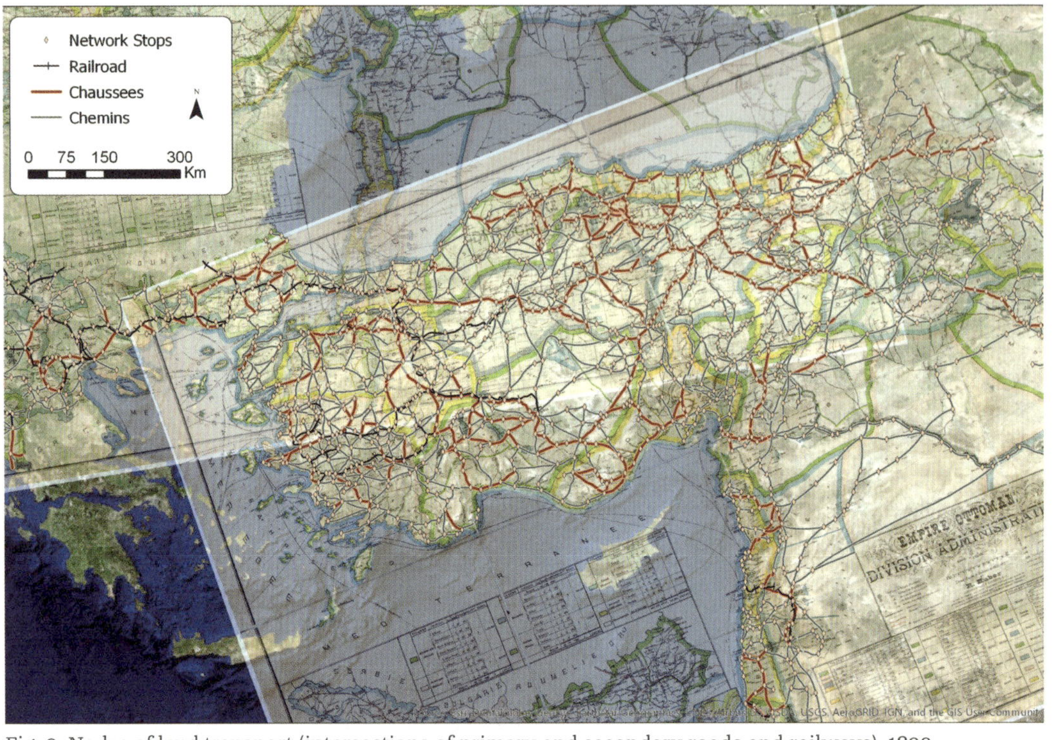

Fig. 9. Nodes of land transport (intersections of primary and secondary roads and railways), 1899.

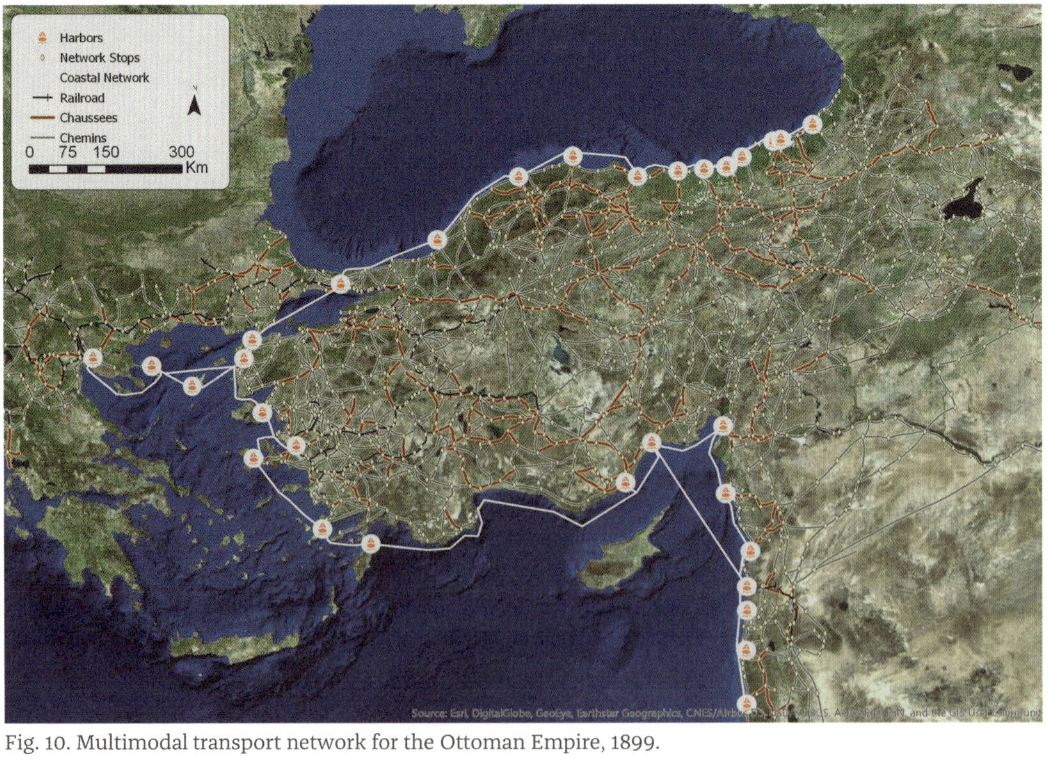

Fig. 10. Multimodal transport network for the Ottoman Empire, 1899.

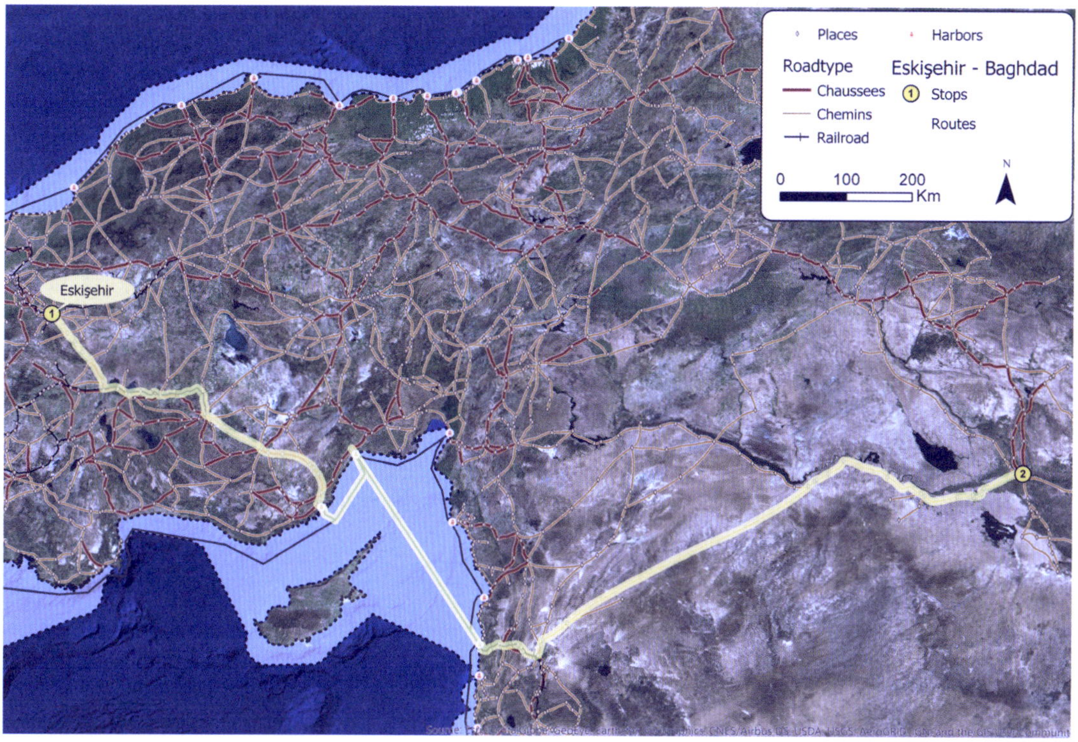

Fig. 11. Hypothetical journey from Eskişehir to Baghdad on the network.

century. This is the first step of an ongoing project. Over time, we aim to develop the model by incorporating additional historical maps, as well as other time, distance, and topographic data. Ottoman provincial statistical yearbooks (*vilayet salnameleri*) from the second half of the nineteenth century are promising sources for travel times in hours. Numerous historical maps provide similar data. A major shortcoming of the current model is that it fails to adjust times and distances according to seasonality, and thus accessibility. To the best of our understanding, this type of information could be gleaned only from careful study of travelogues. Enriching the model with data from the logbooks of additional steamers and other seagoing vessels, as well as the railway timetables from several decades of the nineteenth century, will increase its detail and accuracy with real-life data. Finally, the model presented here makes no use of topographic data. We remain convinced that with careful integration of digital elevation models (DEMs), our GIS-based analyses and simulations will be far more versatile and realistic, able to consider directionality and altitude changes. For a very first step, see our model of an interactive online DEM (UrbanOccupationsOETR, n.d.). These and other future aims should make the model increasingly reliable, and thus increasingly useful as a research tool. We look forward to sharing future iterations of the model as they progress on our project webpage (UrbanOccupationsOETR, n.d.). Sustaining the interactive online presence of digital humanities projects going beyond the lifetime of their funding is a crucial aspect for the future of digital scholarship. To reach this aim we need to go beyond the capabilities of digital repositories for conventional publications or digital applications. We are convinced that ensuring accessibility to academic output in digital formats that allow interactivity on the web is the way forward. We need to establish and maintain interconnected hubs and/or portals for digital scholarship.

Acknowledgments

This work was carried out within the framework of UrbanOccupationsOETR, Industrialisation and Urban Growth from the mid-nineteenth century Ottoman Empire to Contemporary Turkey in a Comparative Perspective, 1850–2000, funded by the European Research Council (ERC) under the European Union's Horizon 2020 research and innovation programme (grant agreement No. 679097). We would like to thank the Director of the Cambridge Group for the History of Population and Social Structure (CAMPOP), Leigh Shaw-Taylor, and his team for our ongoing fruitful collaboration.

Works Cited

Bayram, Selahattin. 2009. "Osmanlı Döneminde Selanik Limanı (Salonica Harbor in Ottoman Era): 1869–1912." PhD diss., Istanbul University.

CAMPOP. n.d. "Urbanization and Economic Development in England and Wales c. 1670–1911." Accessed February 26, 2019. https://www.campop.geog.cam.ac.uk/research/projects/transport/.

Coursera. n.d. "Maps and Geospatial Revolution." Accessed February 26, 2019. https://www.coursera.org/learn/geospatial.

Dobson, Jerome E. 2004. "The GIS Revolution in Science and Society." In *Geography and Technology*, edited by Stanley D. Brunn, Susan L. Cutter, and J. W. Jr. Harrington, 573–87. Dordrecht: Springer.

EarthWorks. n.d. Stanford Libraries: EarthWorks. Accessed February 26, 2019. https://earthworks.stanford.edu/

EMODnet. n.d. European Marine Observation and Data Network. Accessed February 26, 2019. http://www.emodnet.eu/.

Fener Risâlesi. 1911. *Bahr-i Sefîd, Bahr-i Siyah ve Marmara ve Boğazları, Bahr-i Azak, Bahr-i Ahmer, Bâb ei-Mendeb ve Arabistan'ın Cenub-ı Sâhili ile Bahr-i Umman ve Basra Körfezi Fenerlerinin Ta'rifatını Şâmil Risâledir. (Erkan-i Harbiye-i Bahriye Dairesi Ma'rifetiyle Tercüme Edilmişdir)*. Istanbul: Matbaa-i Bahriye.

Gratien, Chris, Michael Połczyński, and Nir Shafir. 2014. "Digital Frontiers of Ottoman Studies." *Journal of the Ottoman and Turkish Studies Association* 1, nos. 1–2: 37–51.

Gregory, Ian, Donald A. DeBats, and Donald Lafreniere, eds. 2018. *The Routledge Companion to Spatial History*. Abingdon: Routledge.

Gregory, Ian N., and Paul S. Ell. 2007. *Historical GIS Technologies, Methodologies, and Scholarship*. Cambridge: Cambridge University Press.

Hadjikyriacou, Antonis, and Elias Kolovos. 2015. "Rural Economies and Digital Humanities: Prospects and Challenges." In *Ottoman Rural Societies and Economies*, edited by Elias Kolovos, 415–21. Rethymno: Crete University Press.

Huber, R., cartographer. 1899. *Empire Ottoman: Division Administrative* [dressée d'après le Salname 1899]. 62 × 44 cm. 1:150,000 scale. Bibliothèque nationale de France, GED-4181. https://gallica.bnf.fr/ark:/12148/btv1b530068545.

Martí-Henneberg, Jordi. 2011. "Geographical Information Systems and the Study of History." *Journal of Interdisciplinary History* 42 (1): 1–13.

MHATNF (Maritime History Archives of Turkish Naval Forces), Logbook Collection, no. 4118.

OpenOttoman. n.d. "A Developing Transnational Digital Collaborative for Scholarship and Public History Focused on the Ottoman World." Accessed February 26, 2019. https://openottoman.org/.

ORBIS. n.d. "The Stanford Geospatial Network Model of the Roman World." Accessed February 26, 2019. http://orbis.stanford.edu/.

Ottoman Railway Company. n.d. "Ottoman Railway Company (Smyrna to Aidin) of His Imperial Majesty the Sultan. Passenger Time Table on and After 29 August 1875." Levantine Heritage. Accessed February 26, 2019. http://www.levantineheritage.com/rail.htm.

Redhouse, James W. 1890. *A Turkish and English Lexicon*. Constantinople: A. H. Boyaciyan.

Sami, Şemseddin. 1901. *Kamus-i Fransevi: Fransızca'dan Türkçeye lügat kitabı = dictionnaire Français-Turc illustre de 3000 gravures*. Istanbul: Mihran.

Stanev, Kaloyan. 2013. "A Historical GIS Approach to Studying the Evolution of the Railway and Urban Networks: The Balkans, 1870–2001." *Historical Methods* 46 (3): 192–201. https://doi.org/10.1080/01615440.2013.804789.

UrbanOccupationsOETR. n.d. "Ottoman Road and Coastal Network 1900." ESRI. Accessed February 26, 2019. http://arcg.is/2hTVeai.

UrbanOccupationsOETR. n.d. "Urban Occupations. Bringing Ottoman/Turkish History into Digital Humanities." Accessed February 26, 2019. https://urbanoccupations.ku.edu.tr/.

Vogel, C., and B. Domann, cartographers. 1908. "Die Balkan-Halbinsel in 4 Blättern: Türkei, Rumänien, Griechenland, Serbien, Montenegro, Bulgarien." 1:1,500,000 scale. *Stieler's Hand-Atlas*, Nr. 51,52,53, 54. Gotha: Justus Perthes. Accessed February 26, 2019. https://earthworks.stanford.edu/catalog/harvard-g6800-1908-v6.

Weiwei, Zhang, and John R. Logan. 2017. "Historical GIS." Oxford Bibliographies. Accessed February 26, 2019. http://www.oxfordbibliographies.com/view/document/obo-9780199874002/obo-9780199874002-0163.xml.

Mapping the *Bostancıbaşı* (Chief Gardener) Registers (1800–1803): An Exploratory Study of Socio-Spatial Differentiation on the Shorelines of Early Modern Istanbul

H. Murat Güvenç

Ayşe Nur Akdal

Introduction

This article is based on a scientific research project titled "Socio-Spatial Structure of the Shorelines of the Golden Horn and the Bosphorus in the Pre-Industrial Era: A Numerical Analysis of the *Bostancıbaşı* (Chief Gardener) Registers on the reign of Selim III (1789-1808) and Mahmud II (1808-1839)," completed in February 2019 at Kadir Has University. We intend to share the project with the public via a printed book and a website by the end of 2020. Designed in line with an exploratory approach, as far as Ottoman historiography is concerned, our research sought mainly to reevaluate—or rather rediscover—two *bostancıbaşı* registers dating to the reigns of Selim III (1789-1808) and Mahmud II (1808-1839) through a new analytical method allowing for spatialization by mapping. The research itself was a first in the field thanks to its data-driven character, questioning a number of postulates in Ottoman historiography. Various maps of the socio-spatial structure of Istanbul's shorelines in the aforementioned times have been drawn up regarding different aspects (Land Use Profile I-II, Occupations I-II, Ethnic-Religious Profiles, Gender, Titles) as a result of analyses conducted (e.g., simple correspondence analysis, cluster analysis). This article briefly introduces the *bostancıbaşı* registers, the methods employed in various stages of research, and the morphological properties of the shoreline, before going on to provide a general perspective on the background of Istanbul's shorelines based on itineraries and secondary sources. The chapter then considers a sample dataset and map derived from a *bostancıbaşı* register dating to the reign of Selim III to present some partial results of the broader research, namely the ethnic-religious profiles. In the final section we discuss the various advantages and disadvantages of implementing methods drawn from the digital humanities for historical research.

About the *Bostancıbaşı* Registers

The *bostancıbaşı* (chief gardener) of the palace was responsible for drawing up a register. According to Bardakçı (2013), the *bostancıbaşı* not only led "a security post" (*Bostancı Bölüğü*) but was also responsible for ensuring public order around the Bosphorus and the Princes' Islands. He was a sort of "harbor master" and a member of the commission authorized to issue construction permits for new buildings on the shorelines of Istanbul. The *bostancıbaşı* was also expected to accompany the sultan when traveling on the imperial caique.

Various speculations offer clues about the original purpose of the *bostancıbaşı* registers. Some argue they were kept to provide information for answering the sultan's question about buildings and their owners when out on the imperial caique, visiting the Bosphorus or the Golden Horn. Others say that the *bostancıbaşı* kept a record of developments on the shorelines under his own authority (Bardakçı 2013, 7–9).

Different collections boast a variety of *bostancıbaşı* registers. Bardakçı cites nine registers known today, while only four of them have been transcribed from Ottoman Turkish for publication (Rado 1972; Koçu 1958a; Kayra and Üyepazarcı 1992; Bardakçı 2013). Bardakçı writes that, apart from the above publications, four undergraduate honors projects on *bostancıbaşı* registers were completed at Istanbul University, Faculty of Letters, Department of History. We found three of these projects (Başaran 1951; Sertel 1951; Süer 1951)—each focused on reproducing the *bostancıbaşı* registers. They contributed to the biographies of important people mentioned in the registers and speculated about the time periods the registers might have been originally drawn up, based on details concerning the years in which these important people held the office and died.

Although several books address certain aspects of parts of the Istanbul shoreline, such as architectural history and literature, no source provides as comprehensive and systematic a dataset for a comparative analysis of the entire shoreline as the *bostancıbaşı* registers. Certain parts of the shorelines covered by these aforementioned sources remain overlooked too. Despite many articles written about it, the Istanbul shoreline has fallen short of being the center of a comprehensive and systematic analysis so far and therefore remains somewhat "uncharted territory." The records from the registers shed light on land use and former residents, hence providing information on the related social and demographic properties of these sites (e.g., ethnic-religious profiles, occupations, genders, and titles). The fact that the buildings along the Bosphorus and Golden Horn shorelines are recorded in a sequential fashion makes these registers suitable for mapping and evaluating, in line with new perspectives and techniques developed in digital humanities. The original layout of the registers begins from the southern shoreline of the Golden Horn, around Topkapı Palace, and continues along the southern shoreline, reaching Eyüp before extending to Tophane along the northern shoreline. Records regarding the Bosphorus usually start from around Tophane on the European side, stretch up to Rumeli Kavağı, and then move on to the opposite shoreline, beginning from Anadolu Kavağı and going all the way down to Kadıköy (Fig. 1).

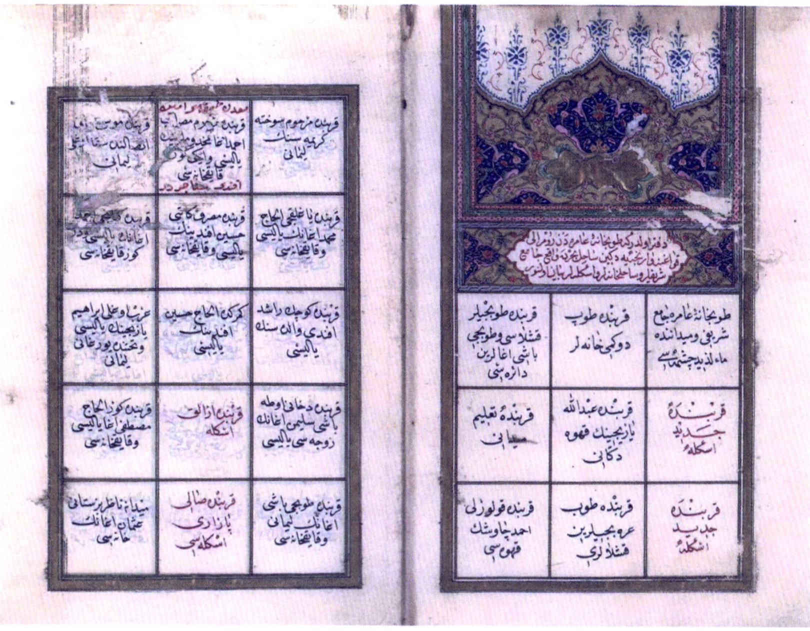

Fig. 1. A transcribed page from the *bostancıbaşı* register of the reign of Selim III (1789–1808).

3.a ١٥

Kurbünde Musa Hanı ittisâlinde Sakaoğlu Limanı	Kurbünde merhûm Musahib Ahmed Ağa mahdûmlarının yalısı ve iki göz kayıkhânesi *Maden Halifesi Emin Efendi müstecîrdir*	Kurbünde merhûm Suhte kerîmesinin limanı
Kurbünde Kebapçı Ahmed Ağa'nın yalısı ve dört göz kayıkhânesi	Kurbünde Masraf kâtibi Hüseyin Efendi'nin yalısı ve kayıkhânesi	Kurbünde Yağlıkçı el-Hâc Mehemmed Ağa'nın yalısı ve kayıkhânesi
Araboğlu İbrahim Yazıcı'nın yalısı ve tahtında Yorganî Limanı	Gümrükte el-Hâc Hüseyin Efendi'nin yalısı	Kurbünde Küçük Râşid Efendi vâlidesinin yalısı
Kurbünde kör el-Hâc Mustafa Ağa yalısı ve kayıkhânesi	*Kurbünde aralık iskele*	Kurbünde Duhânî Odabaşı Süleyman Ağa'nın zevcesi yalısı
Meydâna nâzır Bezistânî Osman Ağa'nın hânesi	*Kurbünde Salıpazarı İskelesi*	Kurbünde Topçubaşı Ağa'nın limanı ve kayıkhânesi

Defter oldur ki Tophâne-i Âmire'den Rumeli Kavağı'na varıncaya değin sâhil-i bahrda vâki' câmî-i şerifler ve sâhilhâneler ve iskeleler beyân olunur

2 *2.b*

Kurbünde topçular kışlası ve topçubaşı ağaların dâiresi	Kurbünde top dökümü hâneleri	Tophâne-i Âmire Câmi-i Şerîfi ve meydânında mâ-i lezîz çeşmesi
Kurbünde tâlim meydânı	Kurbünde Abdullah Yazıcı'nın kahve dükkânı	*Kurbünde cedîd iskele*
Kurbünde Kolozlu Ahmed Çavuş'un kahvesi	Kurbünde top arabacıların kışlaları	*Kurbünde cedîd iskele*

Most data from the Ottoman records regarding places are mainly descriptively written records, making it rather difficult to analyze, evaluate, and compare the texture and characteristics of Ottoman cities. For instance, building and site descriptions from the registers of *qadi*s have yet to be visualized, although they offer rich details regarding the spatial properties of cities. The *bostancıbaşı* registers, which are basically written documents regarding places, have also shared a similar fate. Despite offering tremendous data regarding the socio-spatial aspects of the Istanbul shoreline in the pre-modern period, a comprehensive and systematic evaluation project has yet to be conducted on the *bostancıbaşı* registers until now. The reason for this is likely to be methodological problems about verbal/qualitative data representation rather than any lack of interest in the registers. In that respect, books published on the registers stress that transcribing them is insufficient and that comprehensive, detailed, comparative analyses are required to assess the information at hand adequately (Bardakçı 2013). Difficulties involved with evaluating the qualitative data in a systematic and statistical way have so far confined the evaluative work on the Ottoman archives to the limits of traditional historiography. However, the digital humanities offer interesting research opportunities for reevaluating and reinterpreting such data, which have been treated in a rather superficial manner into present times. We outline the stages of our research, representing a first for Ottoman historiography, in the following section.

Research Stages

Data Classification

Sequential records in the *bostancıbaşı* registers compose the raw data source for the research (Fig. 1). Either a single function or multiple functions are recorded in every box. Except for a few that provide locations alone (Akıntıburnu, Şeytan Akıntısı, Küçük Bebek, etc.), all other records denote function, such as *yalı* (Ottoman waterfront mansion), coffeehouse, pier, boathouse, fishery, mill, etc. Certain functions are associated also with people, like "the *Yalı* of *dhimmi* Moneylender Abraham's son Ohanes" (*Sarraf Abraham oğlu Ohanes zımminin yalısı*) (*dhimmi* is a general term that refers to non-Muslim subjects). Once what is written in every box has been transferred to a spreadsheet table via optical character recognition (OCR) software, new rows are added for every box that includes more than one function. For example, a box reads "Tophane-i Amire Mosque and the Fountain" (*Tophane-i Amire Camii* Şerifi *ve Ma-i leziz* çeşmesi). This box boasts two functions, a mosque and a fountain. In that case the mosque and fountain functions are recorded in separate rows. The same approach is implemented for cases where more than one building with the same function is listed. For instance, when a register states twenty stores selling lemons in a single box, twenty rows are entered in the spreadsheet and each of the stores is coded separately, therefore expanding the list like an accordion. On the expanding list, information in every row is distributed across related column headings (function, ethnic-religious, occupation, gender, title1, title2, religious title, reference, reference data, etc.).

In the next stage, the functions and operation types for land use in the early nineteenth-century Ottoman society to which the *bostancıbaşı* registers date are coded, along with the occupation, status, title, ethnic-religious identity, and other categories of the row, all tied to a key code developed for the project. As a result, entries in every column of the spreadsheet are grouped according to their similarities. In the pre-modern era, some of these functions were not as spatially segregated as today. For example, trade and manufacture coexisted in several cases. Also, coffeehouses were used for accommodation purposes as well (Kırlı and Başaran 2011, 15). To circumvent this problem, land-use profiles are grouped in two clusters: first, on a general level and then on a detailed level. Land use is broken into seven groups in general and coded accordingly as follows: Housing; Gardening & Fishing; Manufacture; Military; Wholesale & Retail Trade, Food & Drink, Accommodation; Transport, Storage & Customs; and Social Services. These seven groups are then subdivided into a total of thirty-one subgroups, each assigned a separate code. For instance, under the Housing group are coded subgroups for House, Pavilion & Palace, Kiosk, *Yalı*, and land.

Needless to say, it is difficult to group the occupations and the occupational titles recorded on a register from the pre-modern period and, in some cases, initiative has to be taken. Although few in number, people with two occupations are listed in the registers: for instance, the "House of Hüseyin Beşe, the boatman and *imam*" (*Kayıkçı İmam Hüseyin Beşe'nin Hanesi*). In some cases, owning property is confused with having an occupation, such as in the case of the "Rice-Merchant İsmail Ağa's Jewish House and Coffeehouse" (*Pirinççi İsmail Ağa'nın yahudhanesi ve kahvesi*). In that case, İsmail Ağa's occupation is assumed to have been rice seller. Of course, the very same person could also have run the coffeehouse and the Jewish House, too. Nevertheless, no speculations are made because the registers do not provide clear information on the matter. The registers usually also state the occupations of people living in the houses, like the "*Yalı* of *dhimmi* Tanaş the furrier" (*Kürkçü Tanaş zımminin yalısı*). The occupations of workplace owners are usually not mentioned, such as in "Mehmet Efendi's Coffeehouse" (*Mehmet Efendi'nin Kahvesi*). Because the registers provide us with information only on actual residents, as opposed to storekeepers, etc., we are able to comment only on the occupations of the resident population. Of course, in the example above, "*Mehmet Efendi*" is likely to be both the property owner and the manager of the coffeehouse. Nonetheless, we have refrained from recording this speculation because the register remains ambiguous. After all, it would matter little if a function stated as "coffeehouse" on the land-use map was recorded also with its coffeehouse owner on the occupational map. The registers rarely explicitly state both the owner of a workplace and the function of the workplace; for instance, we do not see entries like "Coffeehouse Keeper Mehmet Efendi's Coffeehouse" (*Kahveci Mehmet Efendi'nin Kahvesi*). In this case, because Mehmet Efendi is specifically stated to be a coffeehouse keeper in the register, his occupation is taken into consideration as a coffeehouse keeper.

Occupational profiles are grouped on three levels, from general to detailed. On the first level grouping, titles include tradesmen and merchants, statesmen, and foreign diplomats and dragomans. Because the second and third level groupings are much more detailed, they will not be discussed in this chapter. Originally developed by the Cambridge Group for the History of Population and Social Structure (CAMPOP), the Primary, Secondary, Tertiary System (PST) of occupational coding categories have been used for grouping some of the occupations recorded by the register. The wide variety of statesmen of different levels living on the Istanbul shoreline is grouped through the very key code we developed. Apart from that, grouping other profiles (e.g., ethnic-religious structure, gender, titles) and coding them is not as difficult and detailed as it is for land use and occupations.

Although information from registers of the pre-modern period is not always eligible for classification in modern senses, the special key code we developed managed to group the information about land use and social structure without compromising their authenticity. Although originally drawn up in a considerably systematic fashion, the *bostancıbaşı* registers also present challenges that are impossible to resolve. For instance, when a register record refers to streets full of coffeehouses or stores of the same kind, it is unclear how many coffeehouse or store functions should be recorded. Whenever such plural expressions are used for the same functions, we have recorded them at least twice. Additionally, another problem is the "weight" of the functions, as far as the places are concerned. The importance of the functions can vary depending on different aspects, such as the number of people working or living inside the place, or the production and sales capacity, etc. For example, a vegetable market (*sebzehane*) and a greengrocer's have different function weights in every aspect. In the same way, Çırağan Yalısı and other *yalıs* are completely different from each other, as are single houses when compared to other entities that host different numbers of people (e.g., rooms for single men [*bekar odası*]; a residential inn [*han*]; a Jewish House [*yahudhane*], which was a kind of hostel built and run by Jewish community members or Muslim entrepreneurs that had numerous rooms intended for Jewish paupers to rest or reside). Unfortunately, the lack of data makes it impossible for researchers to account for such qualitative differences in quantitative terms.

GIS Coordination

Drawn up on a parcel-by-parcel basis, the *bostancıbaşı* registers are actually not suitable for direct geographic analysis. For that purpose, we divided Istanbul's shoreline into units of different lengths, called "segments," involving varying amounts of data (Fig. 2). This dividing operation was conducted in three ways:

1. Using landmarks (such as mosques, piers, etc.) shown on the *bostancıbaşı* registers that provide geographic positions
2. Using abrupt changes in records of ethnic-religious structures (for instance, if ethnic-religious profiles located between landmarks differ significantly, the segment is subdivided)
3. Using abrupt changes in records of land use (for instance, if land-use functions between landmarks differ significantly, the segment is subdivided)

Various cartographic sources from different times were used to determine the

SEGMENTS LOCATED ON WESTERN SHORELINE OF THE BOSPHORUS

1001 Tophane-Salıpazarı
1002 Salıpazarı-Fındıklı
1003 Fındıklı-Kabataş
1004 Kabataş-Karabali
1005 Karabali-Hayrettin İsk.
1006 Hayrettin İsk.-Beşiktaş İsk.
1007 Beşiktaş İsk.-Çırağan Sarayı
1008 Çırağan Sarayı
1009 Çırağan-Yahya Efendi Deresi
1010 Yahya Efendi Deresi-Ortaköy
1011 Ortaköy-Ortaköy Camii
1012 Ortaköy Camii-Koyun İsk.
1013 Koyun İsk.-Defterdar Burnu
1014 Defterdar Burnu-Ekmekçioğlu Deresi
1015 Ekmekçioğlu Deresi-Kuruçeşme İsk.
1016 Kuruçeşme İsk.-Bostani Ocağı
1017 Kuruçeşme Bostani Ocağı
1018 Kuruçeşme Bostani Ocağı-Arnavudköy İsk.
1019 Arnavudköy İsk.-Akıntıburnu
1020 Akıntıburnu-Bebek
1021 Bebek-Küçük Bebek
1022 Küçük Bebek-Kayalar
1023 Kayalar
1024 Rumelihisarı-Mahmut Han Camii
1025 Rumelihisarı-Şeytan Akıntısı
1026 Şeytan Akıntısı-Baltalimanı
1027 Baltalimanı-Mirgunoğlu
1028 Mirgunoğlu-Tokmakburnu
1029 Tokmakburnu-İstinye İsk.
1030 İstinye İsk.-Mezarburnu İsk.
1031 Mezarburnu İsk.-Pazarkayığı İsk.
1032 Pazarkayığı İsk.-Ekmekci İsk.
1033 Ekmekci İsk.-Köy İsk.
1034 Köy İsk.-Köybaşı (Yeniköy) İsk.
1035 Köybaşı (Yeniköy) İsk.-Kalender Bahçesi
1036 Kalender Bahçesi
1037 Tarabya Sınırı-Tarabya Mescidi
1038 Tarabya Mescidi-Tarabya İsk.
1039 Tarabya İsk.-Kireçburnu
1040 Kireçburnu
1041 Kefeli Karyesi
1042 Büyükdere Çayırı
1043 (Büyükdere_1) Büyükdere Çayırı-Mehmet Paşa Camii
1044 (Büyükdere_2) Mehmet Paşa Camii-Mezarburnu (Sarıyer Sınırı)
1045 Mezarburnu (Sarıyer Sınırı)
1046 Mezarburnu-Sarıyer İsk.
1047 Sarıyer İsk.-Yenimahalle
1048 Yenimahalle
1049 Tabya
1050 Rumelikavağı

SEGMENTS LOCATED ON EASTERN SHORELINE OF THE BOSPHORUS

2001 Anadolukavağı
2002 Yuşa Tabyası
2003 Umuryeri
2004 Serviburnu
2005 Değirmenlik
2006 Hünkar İsk.
2007 Hünkar İsk.-Yalıköy İsk.
2008 Yalıköy İsk.-Beykoz İsk.
2009 Beykoz İsk.-Sultaniye
2010 Sultaniye Bahçesi
2011 İncirköy İsk.-Paşabahçe İsk.
2012 Paşabahçe İsk.-Çubuklu Bahçesi
2013 Çubuklu Bahçesi
2014 Çubuklu Bahçesi-Kanlıca İsk.
2015 Kanlıca İsk.-Bahai Körfezi
2016 Bahai Körfezi-Anadolu Hisarı
2017 Anadoluhisarı
2018 Göksu Deresi
2019 Göksu
2020 Göksu-Kandilli Camii
2021 Kandilli Camii-Şattiler (Akıntıburnu)
2022 Şattiler (Akıntıburnu)-Vani Efendi Camii
2023 Vani Efendi Camii-Kulebahçe
2024 Kulebahçe
2025 Kulebahçe_2
2026 Kulebahçe-Çengel Karyesi İsk.
2027 Çengel Karyesi İsk.-Abdulhamid Camii
2028 Abdulhamid Camii-İstavros İsk.
2029 İstavros İsk.-Nakkaş Paşa
2030 Nakkaş Paşa-Kuzguncuk İsk.
2031 Kuzguncuk İsk-
2032 Kuzguncuk-Öküz Limanı
2033 Öküz Limanı
2034 (Üsküdar_1) Öküz Limanı-Mihrimah Sultan Camii
2035 (Üsküdar_2) Mihrimah Sultan Camii-Şemsi Paşa Camii
2036 (Üsküdar_3) Şemsi Paşa Camii-Fatih Sultan Mehmet Han Camii
2037 (Uskudar_4) Fatih Sultan Mehmet Han Camii-Harem İsk.
2038 Harem
2039 Kavak İsk.
2040 Haydarpasa
2041 Haydarpasa-Kadıköy İsk.
2042 Kadıköy İsk.-Moda Burnu
2043 Fenerbahçe

SEGMENTS LOCATED ON SOUTHERN SHORELINE OF THE GOLDEN HORN

3101 Yalı Kasrı-Ocaklar
3102 Ocaklar-Yeşil Kiremitli Camii
3103 Yeşil Kiremitli Camii-Kuşcubaşı Camii
3104 Kuşcubaşı Camii-Selim Paşa Namazgahı
3105 Selim Paşa Namazgahı-İstanbul Ağası İsk.
3106 İstanbul Ağası İsk.-Gümrük-ü Kebir İsk.
3107 Gümrük-ü Kebir İsk.-İzzet Paşa Camii
3108 İzzet Paşa Camii-Duhan Gümrüğü İsk.
3109 Duhan Gümrüğü İsk.-Manavlar
3110 Manavlar-Çardak İsk.
3111 Çardak İsk.-Odunkapısı İsk.
3112 Odunkapısı İsk.-Ayazmakapısı İsk.
3113 Ayazmakapısı İsk.-Tekfurdağı İsk.
3114 Tekfurdağı İsk.-Tahtapuş İsk.
3115 Tahtapuş İsk.-Unkapanı İsk.
3116 Unkapanı İsk.-Balıkhane İsk.
3117 Balıkhane İsk.-Tüfenkhane İsk.
3118 Tüfenkhane İsk.-Cibali İsk.
3119 Cibali İsk.-Ayakapısı İsk.
3120 Ayakapısı İsk.-Yenikapı İsk.
3121 Yenikapı İsk.-Petrokapısı İsk.
3122 Petrokapısı İsk.-Fenerkapısı İsk.
3123 Fenerkapısı İsk.-Balat İsk.
3124 Balat İsk.-Kalafatçılar Meydanı
3125 Kalafatçılar Meydanı-Ayvansaray İsk.
3126 Ayvansaray İsk.-Yavedud İsk.
3127 Defterdar İsk.-Eyüp İsk.
3128 Eyüp İsk.-Çukur Saray
3129 Çukur Saray mahali

SEGMENTS LOCATED ON NORTHERN SHORELINE OF THE GOLDEN HORN

3201 Karaağaç Bahçesi-Sütlüce İsk.
3202 Sütlüce İsk.-Kumbaracı Camii
3203 Kumbaracı Camii-Buzhane İsk.
3204 Buzhane İsk.-Piri Mehmet Paşa İsk.
3205 Piri Mehmet Paşa İsk.-Hamam İsk.
3206 Hamam İsk.-Hasköy İsk.
3207 Hasköy İsk.-Fatih Sultan Han Camii
3208 Aynalıkavak Sarayı ve Tersane Bahçesi
3209 Tersane-i Amire
3210 Meyyit İsk.-Azapkapı İsk.
3211 Azapkapı İsk.-Kürkçükapısı İsk.
3212 Kürkçükapısı İsk.-Yağkapanı İsk.
3213 Yağkapanı İsk.-Karaköy İsk.
3214 Karaköy İsk.-Galata Gümrüğü İsk.
3215 Galata Gümrüğü İsk.-Kurşunlu Mahzen
3216 Kurşunlu Mahzen-Mumhane İsk.
3217 Mumhane İsk.-Eğrikapı İsk.
3218 Eğrikapı İsk.-Kireçkapı İsk.
3219 Kireçkapı İsk.-Debbağhane İsk.
3220 Debbağhane İsk.-Dükkanlar

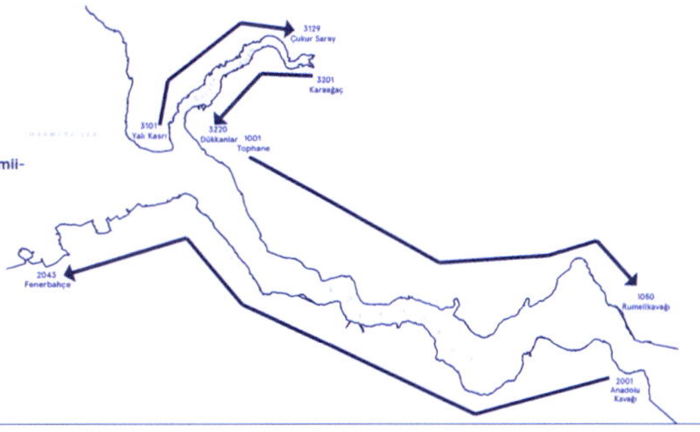

Fig. 2. Segments located on shorelines of Istanbul.

unsettled areas and landmarks properly. The Pervititch maps, German maps, Kauffer maps, Kaminar maps, Ayverdi maps, Sedad Hakkı Eldem's map on Bosphorus *yalıs*, *Rumeli ve Anadolu Ciheti* maps, and finally sections of Müller-Wiener's Istanbul Port Map have been studied and integrated into a Geographic Information System (GIS). Subsequently, the shoreline location was determined as accurately as possible considering the original time in which the landmarks were marked on the registers so that the segments accord to the abovementioned stages.

Istanbul's shoreline from the *bostancıbaşı* register of Selim III is divided into 142 segments, while the shoreline from the *bostancıbaşı* register of Mahmud II is divided into 140 segments. The segments on both registers have been coordinated and hence allow comparison. While the *bostancıbaşı* register of Mahmud II ends at Haydarpaşa, the register of Selim III stretches as far as Fenerbahçe and, accordingly, has three more segments. The high number of segments and the consistency of their recording increases the sensitivity of the maps and the resulting analyses.

In the key-code system, which we established to distinguish the segments, codes starting with "1000" represent the west shoreline of the Bosphorus, codes starting with "2000" represent the east shoreline of the Bosphorus, codes starting with "3100" represent the southern shoreline of the Golden Horn and, finally, codes starting with "3200" represent the northern shoreline of the Golden Horn. Excluding all unsettled areas, the Istanbul shoreline is divided into 142 segments: 1001–1050 covers the west side of the Bosphorus, from Tophane to Rumeli Kavağı; 2001–2043 covers the east side of the Bosphorus, from Anadolukavağı to Fenerbahçe; 3101–3129 covers the south side of the Golden Horn, from Yalı Kasrı to Çukur Saray; and 3201–3220 covers the north side of the Golden Horn, from Karaağaç Bahçesi to Debbağhane İskelesi. Because the segments are usually sequential, the end of one segment corresponds to the beginning of another. For this reason, they are named according to the places at their start and end points, such as Tophane-Salıpazarı, Salıpazarı-Fındıklı, Fındıklı-Kabataş, etc. Nonetheless, certain segments have turned out not to be sequential, and they are named in singular fashion, such as Kayalar, Kireçburnu, Kefeli Karyesi, and Tabya (Fig. 2).

Method and Analysis

Techniques allowing statistical evaluation of nominal data have been developed only in recent decades. Because it is rather difficult to visualize and interpret observations comprehensively through conventional techniques when the number of observations exceeds fifty, the practical solutions offered by simple and multiple correspondence analyses need to be emphasized. Consequently, the difficulty of analyzing a typical *bostancıbaşı* register, with up to two thousand data entries, through conventional techniques is obvious. This methodological challenge is overcome by introducing the complementary use of correspondence and cluster analyses developed by Ludovic Lebart (1994). In this respect, correspondence analysis is adopted as a pattern recognition instrument, while cluster procedures are employed for achieving categorization with a minimum loss of information. Using the two methods together, our approach eliminates the weaknesses of one methodology through another, thereby leading to an efficient and robust complementary representation and visualization model. The output from the model is thus as suitable for data visualization and mapping in GIS environments as is any quantitative representation. The complementary model introduced by Lebart can easily become functional through recently developed software products, such as XLSTAT. Developed by Lebart, these methods are extremely apt for visualizing and mapping qualitative data. Our project has used an interface developed in the Istanbul Studies Center (ISC), coded in R and made functional by Lebart's model. Called CLUSTERS, this interface is currently used by many projects undertaken in the ISC. CLUSTERS operates with an XLSTAT package featuring macros in C++. The entire pattern recognition, clustering, and legend formation processes, as well as map legend categories, have been completed via CLUSTERS. The Lebart model is implemented within the scope of a two-stage classification. The first stage is a preliminary classification, and the second stage is the final clustering. Sample outputs from the interface will be reviewed in detail below and will consider the results of an analysis on the ethno-religious profile of the region.

Visualization and Mapping

Once the analysis of the data has been conducted, the visualization stage begins. ISC fellow researcher Murat Tülek was responsible for visualizing and designing the analyses undertaken according to the segments of shorelines determined through our GIS.

Historical Background: Changes in Settlement Topography along the Bosphorus Shoreline

The Bosphorus shoreline has been examined through two particular historical periods: the seventeenth century and the eighteenth century. The shoreline was one of the main historical routes followed by travelers and was the subject of numerous itineraries. A selection of such firsthand sources of Armenian travelers/authors, such as Kömürciyan (1637–1695), Hovannesyan (1740–1805), and İnciciyan (1758–1833), has been drawn upon to supplement and further support the data gleaned from the *bostancıbaşı* registers. Accordingly, the urban landscape of the Bosphorus during the two successive historical periods has been presented through a close examination and comparison of sources.

Itineraries, registers, and travelers

As Kafescioğlu (2011, 139) suggests, a study focusing on structural and representative similarities, as well as differences between itineraries, can provide clues about Ottoman mentalities and imaginaries. Of course, one should consider for whom these images were actually produced. In general, such works are accepted to have been written for the literate elites of Ottoman society; this is a significant point for this study, as the travelers/authors examined stand out on the basis of their education and are rather few in number. Having said that, one could speculate that, when examined closely, the sources also seem to have been drawn up for a more diverse readership than the sultan and the elite.

A common feature of itineraries and registers is that, although they start from different locations, they all follow a certain route along Istanbul's shoreline. We do not know how old exactly the practice of traveling and describing the city through a tour along its shoreline was. Kafescioğlu (2011, 140–41) describes how a spatial paradigm conceptualizing the topography primarily as a route shaped both literary and visual texts in the pre-modern era. This spatial perception was not limited to Ottoman or Eastern realms and was also common in medieval Europe; the first known examples are the Christian pilgrimage route maps. Regarding Istanbul, the primary preferred travel route was the Golden Horn and Bosphorus shorelines. One of the best-known examples of this route is found on a map of the mid-seventeenth century, the original version of which is in the Bibliothèque Nationale de Paris. In the map, numerous neighborhoods, the city gates, and significant edifices and piers were meticulously recorded (Orbay 2011, 204–5). This tradition continued into the early nineteenth century with the *bostancıbaşı* registers. The shoreline was commonly described as a route in maps, itineraries, and certain sub-genres of *divan* literature (*sahilname*). As Orbay (2011, 206-207) states, the shorelines were the subject of rich literary images in the same period.

About Travelers

One of the paradigmatic itineraries of Ottoman literature, *Book of Travels of Evliya Çelebi* (*Evliya Çelebi Seyahatnameleri*), was written in line with two main textual criteria: descriptions referred to geographical/spatial aspects, and narratives referred to historical/chronological aspects (Kafescioğlu 2011, 139). Armenian travelers who wrote about the city used the same method, ornamenting their spatial descriptions with myriad narratives and stories. Nevertheless, their narratives were mainly about specific locations or particular elements therein. For instance, in Kömürciyan's description of a Sufi lodge (*mevlevihane*) in Beşiktaş, he mentions the banning of the lodge, the person responsible for the ban (Vanî the preacher), and how he demanded the extermination of "infidels" after the Ottoman defeat in Austria (*Nemçe*) (Kömürciyan 1988). In Hovannesyan's description of the Çırağan Yalısı, he immediately talks about the Tulip Era (*Lale Devri*), celebrations organized by Grand Vizier İbrahim Pasha (*Sadrazam İbrahim Paşa*), and the lantern decorations (*kandil*) placed around the city (Hovannesyan 1996, 46). At times, these travelers also relate "social functions" fulfilled by these locations. For instance, Kömürciyan writes about the Göksu Creek (*Göksu Deresi*) near Anadolu Hisarı, across from Rumeli Hisarı, and narrates how locals visited the cherry gardens on the seaside in a festive spirit in the spring (Kömürciyan 1988, 42). Unlike the itineraries, the *bostancıbaşı* registers focus more on functions and property, whereas they give no information about certain topics, such as non-Muslim houses of worship, cemeteries, etc. The narratives presented in itineraries are absent in this genre. Again, unlike the itineraries, the *bostancıbaşı* registers follow a consistent route along the shoreline and focus only on the buildings or other items on the waterfront. Itineraries, on the other hand, sometimes get into the details of the inland.

The Bosphorus Shoreline in the Seventeenth and Eighteenth Centuries

> *Nobles look for a place most of the time,*
> *Unknown to many and far away from people,*
> *Turkish, Greek and Armenian elites.*
> *Garnished the shores with magnificent edifices.*
> —G. V. İnciciyan, *Boğaziçi Sayfiyeleri*

Was the eighteenth century indeed a period of change? If so, how should we interpret these changes? According to Eldem (1999), the eighteenth century does not fit into a particular category. Based on this approach, the period can be classified as an age of regression or an age of modernization/Westernization. A significant transformation was occurring in Europe, and the Ottoman state, too, was going through "different times." However, one should not overestimate the level of change in the latter. Reforms implemented in the Ottoman realm at this time were essentially a search within the existing order for solutions to problems stemming from the "outside." This is particularly important because neither a questioning and redefinition of political organization nor a rethinking of the relations between political elites and the society were among the legitimate solutions to which the Ottoman state appealed (Eldem 1999, 197). One particular change in the sociopolitical scene of the period was that elites were becoming more visible in public through their buildings, artifacts, and festivals.

Hamadeh (2007), like Eldem, suggests that changes throughout the eighteenth century cannot be reduced to "Westernization," although there was certainly some European influence and inspiration. According to both authors, this period had particular intrinsic dynamics. One of the spatial manifestations of these dynamics was the expansion of the Palace toward the peripheries, owing to increases in wealth, while another was the expansion of the military class across new neighborhoods of the capital (Eldem 1996, 20; Hamadeh 2007, 338). Indeed, Hovannesyan's manuscript mentions the Tophane and Haydarpaşa neighborhoods together with the novel military barracks and garrisons. The 1801-1803 *bostancıbaşı* registers also record military officers among the significant residents of these neighborhoods, such as *silahtar ağası*, *topçubaşı*, and *çavuşbaşı*.

Hamadeh's (2007) narrative opens with Sultan Ahmed III (r. 1703-1736), who spent most of his early years outside the city, finally coming back to Istanbul in the eighteenth century—a period marked by the most rigorous urban-development and construction wave since the colossal reconstruction projects ordered by Mehmed II, following the city's conquest. Ruling elites and the Palace undertook building projects, including palaces, masjids, mosques, fountains, roads, and water networks (Hamadeh 2007, 15). These projects focused partially on the vicinity of the Golden Horn and mainly on the Bosphorus shoreline. The reputation of the members of the ruling elite expanded to even the remotest villages of the Bosphorus through the buildings of fountains or *yali*s. Not every corner of the Bosphorus attracted the same level of demand, however. For instance, İnciciyan (2000, 167) states that Anadolu Hisarı was not a particularly popular place for recreational areas; therefore, it was not "colonized," unlike other, more popular areas. This resulted from the military function of the neighborhood and its lack of a northern breeze, which, he argues, was essential for cooling.

The visibility of the ordinary people and the urban poor rose during the eighteenth century as a consequence of public works and improvements undertaken in the city (Artan 2010, 308). Similarly, the presence of the urban middle classes became more prominent, and they started to influence the literary culture of the period (Hamadeh 2007, 19). Possibly owing to status concerns, the middle classes sponsored building projects as well (Hamadeh 2007, 30). It is no doubt that owning a *yalı* on the Bosphorus shoreline was a clear sign of prestige and status. For instance, one of the first *yalıs* built on the Bebek shoreline in 1720 was owned by Nureddin Ağa. He may have been an ordinary baker but, as the years passed, he managed to win the navy tender for sea biscuit (*peksimet*) provisioning and went on to become an upper-class member of society, showing off his wealth with new buildings. In addition to his *yalı* and sea-biscuit workshop, he sponsored a fountain and a pier house (*liman köşkü*) in Bebek and started a pious foundation (*vakıf*) in his own name. These sorts of public works were reputable and prestigious in the period. As Aydın (2016, 1-5) states, a vertical mobility through posts in the state administration was available for people from various classes in Ottoman society. But what can we say about entrepreneurs? Wealth accumulation for an entrepreneur depended strictly on having "good relationships" with statesmen. Interestingly, in the 1801-1803 *bostancıbaşı* registers, one can see descriptions about wealthy landowners who were not part of the bureaucratic elite. This was probably not the case in the beginning of the eighteenth century, and it seems that the dynamism from the beginning of that century led to a change in the social topography of the shoreline as the years passed.

Following two big construction periods—first, the reconstruction of the city by Mehmed II and, second, the intensive mosque projects of Mimar Sinan—the eighteenth century was an age of expansion along the Bosphorus shoreline, with the construction of new palaces, recreation areas, settlements, and, of course, *yalıs*. Yerasimos (2010) mentions a pattern of expansion from the center toward the periphery in cities conquered by the Ottomans. The eighteenth century was probably the period when this happened most intensively (Yerasimos 2010, 195-209). Since the properties of non-Muslims were confiscated and transferred to Muslim elites, this expansion also meant conquest. Hamadeh (2007, 69) gives numerous examples of such confiscations. Neither İnciciyan nor Hovannesyan, however, mention any event of confiscation. On the contrary, they narrate events of abandonment of property by elites, which they suggest often resulted in moral corruption. For instance, İnciciyan writes about one of the sultan's gardens in Bebek that became a haven for riffraff after its abandonment. The place was zoned for construction later by İbrahim Paşa, the grand vizier of Sultan Ahmed III. The vizier built a palace, a bathhouse, a bazaar, and a mosque in the neighborhood. İnciciyan (2000, 114) describes the developments in a poem:

All four sides are for sale in the district
Both the hill and the shore
Up until the palace of the Caliph
Now everyone is in a race [to buy]

According to Hamadeh (2007, 67), the western shoreline had historically been more populated than its eastern counterpart. The data gathered from the *bostancıbaşı* registers verify these claims. The western side was the most prominent region among urban dwellers outside the city. Hamadeh (2007, 69) states that after the gradual construction of new gardens along the western (*Rumeli*) shoreline of the Bosphorus, some properties of non-Muslim families, especially in the zone between Beşiktaş and Bebek, were confiscated or destroyed, and the dwellers had to move away to other districts. Although confiscation/transferring was a very common practice, such events are rarely noted in itineraries.

Hovannesyan describes some churches close to the seaside, yet no churches or synagogues appear in the 1801-1803 *bostancıbaşı* registers. The reason may be the demolishing of non-Muslim houses of worship and/or that they were deemed too irrelevant to be recorded. For instance, İnciciyan (1976) writes that the foundation of a mosque in Ortaköy led to the demolishing of a Greek church in 1720-1721, and the non-Muslim community was forced to leave (Hamadeh 2007, 70). Not only confiscations, but also legal enactments served to diminish the non-Muslim presence along the Bosphorus shoreline. Throughout the eighteenth century, several laws were enacted that forbade non-Muslims to build in the Tophane, Galata, Beşiktaş, and Ortaköy neighborhoods (Hamadeh 2007, 74). It was particularly the non-Muslim subjects who suffered most from the expansion of the properties of the sultan and the elites along the Bosphorus. Confiscations must have harmed poor people from Bosphorus villages also—those who made their livings by fishing and farming and stayed in their homes all four seasons, visiting Istanbul rarely.

During the eighteenth century, private property became quite widespread (Hamadeh 2007, 67). Also, according to the 1801-1803 *bostancıbaşı* registers, it was common to have more than one residence. This was partly because the right to build palaces in the capital, which had previously been restricted to the sultan, was extended also to the daughters and nieces of Sultan Ahmed III. This led to an opportunity to increase the public visibility of the women of the dynasty (Artan 2010, 305). In these *bostancıbaşı* registers, the female owners of *yalıs* were registered in reference to their husbands, fathers, or brothers. Women also owned houses, Jewish Houses (*yahudhanes*), shops, and land (albeit these items were smaller in number). Artan (2010, 310) notes that lower status elites began to build fountains along the Bosphorus shorelines during this period as well. Nevertheless, water provision and security remained the two main problems of the Bosphorus. Most of the sultan's gardens were described in the registers together with the gardeners' security post (*bostancılar ocağı*) responsible for them.

The Shorelines of Istanbul during the Reign of Selim III (1789-1808): The Results of an Analysis of the Ethno-Religious Profile of the Region

We now turn to the classification, analysis, and mapping completed through the analysis of the names recorded in the *bostancıbaşı* registers from the period of Selim III. As described above, every box in the *bostancıbaşı* registers represents a spatial function. Certain functions can be associated with a property owner or a leaseholder, in line with the way they are defined in the registers. In such cases, in fact, the names of the property owner and the tenant are only rarely omitted. Registers sometimes provide information about occupations, titles, and references too. Learning about the names of property owners or tenants allows us to interpret their ethnic-religious characteristics and genders. Additionally, ethnic-religious identifiers such as "Jewish" and *dhimmi* (see above) make classification easier. While classifying data on ethnic-religious profiles, we first distinguished Muslims, Jews, and *dhimmis*, in line with the definitions used in the registers. Subsequently, people listed as *dhimmis* have been identified as Armenians or Greeks (all Orthodox Christian subjects of the sultan) through onomastics.

We grouped certain embassy employees and merchants as foreigners/*beratlıs*—*beratlı* referring to Ottoman subjects with "berat" licenses to work either for foreign embassies or as merchants. Such individuals include foreign nationals, such as Austrians (migrated from *Nemçe*), British, Russians, and *Frenks* (with *Frenk* used for Europeans in general, and French in particular), or persons working for foreign countries. In addition, we have recorded Boyars, an aristocratic title in Slavic countries, separately because they were recorded separately and exclusively in the registers and were noteworthy in number. In this respect, we relied on and greatly appreciate the counseling of Edhem Eldem via personal communication.

At first glance, the Muslims seem to be dominant in the groupings. Around 60% of the shoreline is composed of Muslims. The analyses take proportional representation into consideration, however, instead of simple majorities. Accordingly, for a place to be shown as predominantly Muslim over other ethnic-religions elements, a numerical superiority is insufficient. In order for a segment to be marked predominantly Muslim, the number of Muslims in that segment needs to be higher than elsewhere along that shoreline. In other words, the Muslim population needs to be overrepresented. We will see the effects of this approach in detail when interpreting the results below.

As a result of analysis of the *bostancıbaşı* registers from the period of Selim III, the 142 segments that compose Istanbul's shoreline were divided into 13 clusters on the basis of ethnic-religious profiles. Under every cluster lies a "segment"—a unit of observation concerning the table adopted so that the table's quantitative data can be represented spatially. For the sake of clear visualization, these 13 clusters, representing ethnic-religious characteristics in different ratios, are grouped under six generalized titles by considering overrepresented elements within each cluster: Predominantly Jewish; Predominately Armenian; Predominantly Armenian & Greek; Predominantly Foreigner/*Beratlı*; Predominantly Greek; and Predominantly Muslim. Two of the clusters are non-residential areas representative of commercial, military, and agricultural functions; these are excluded from evaluation here. The remaining 11 clusters show both overrepresented and underrepresented ethnic-religious elements, as described below according to their six generalized titles.

1. *Predominantly Jewish*

Jews are overrepresented in the clusters in this section. There are 44 people in total in the C1 cluster, accounting for 1.99% of the exemplary shoreline population (Table 1). The first row of each cluster shows the concentration of ethnic-religious groups in it, while the cells highlighted light gray show the overrepresented elements. Jews and Armenians are overrepresented while Greeks, Muslims, and Non-Residential Areas are underrepresented.

Table 1. Jews 65.91% (4.53%) - Armenians 13.64% (3.76%)

TOTAL PEOPLE	**%**	**CLUSTER**			
44	1.99	C1			
ROW 3-1	**Jews**	**Armenians**	**Greeks**	**Non-Residential**	**Muslims**
Concentration Index	19.13	3.38	–1.78	–2.68	–3.77
% Share in Ethno-Religious Group	29	7.23	0.61	0.7	0.12
% Share in the Cluster	65.91	13.64	4.55	13.64	2.27
% Share in the Shoreline Population	4.53	3.76	14.86	38.83	36.66

2030	Nakkaş Paşa-Kuzguncuk İsk.
2031	Kuzguncuk İsk.-
3218	Eğrıkapı İsk.-Kireçkapı İsk.
3219	Kireçkapı İsk.-Debbaghane İsk.

In other words, although those underrepresented are also included in this cluster, they are represented at a much lower rate compared to other clusters they constitute. Any other category not stated in Table 1 is totally unrepresented and considered absent here. The second row of each cluster reflects what percentage of ethnic-religious elements is present in the cluster. According to Table 1, 29% of all Jews, 7.23% of all Armenians, 0.61% of all Greeks, 0.7% of all Non-Residential Areas, and 0.12% of all Muslims are found in this cluster. The third row demonstrates the proportional distribution of the ethnic-religious profile within the cluster. The C1 cluster consists of 65.91% Jews, 13.64% Armenians, 4.55% Greeks, 13.64% Non-Residential Areas, and 2.27% Muslims. Lastly, to put things in perspective, the fourth row indicates the proportional distribution of the elements constituting the ethnic-religious profiles over the entire shoreline. When listed from high to low, these ratios turn out to be 38.83% Non-Residential Areas, 36.66% Muslims, 14.86% Greeks, 4.53% Jews, 3.76% Armenians, 0.95% Foreigners/*Beratlıs*, and 0.41% Boyars. Segments similar to each other in terms of the cluster values are located below the table, which in this case cover the eastern shoreline of the Bosphorus, near Kuzguncuk, and the northern shoreline of the Golden Horn, near Karaköy.

When interpreting Table 1, a comparison can be drawn between the fourth and the third rows. In this respect, Jews constitute 4.53% of the entire exemplary shoreline population, while they are represented by 65.91% in the related segments. Therefore, while the related segments predominantly contain Jews, it is noted that a certain ratio of Armenians and Jews live together.

There are 79 people in total in the C2 cluster, hosting 3.58% of the population (Table 2). Related segments are located in the vicinity of Unkapanı and Balat on the southern shorelines of the Golden Horn. Only Jews are predominant in these segments while other elements are underrepresented or absent.

The C12 cluster represents 34 people, corresponding to 1.54% of population (Table 3). Spatially, the cluster covers two segments located in the vicinities of Unkapanı, in the southern region of the Golden Horn, and Hasköy, in the northern region of the Golden Horn. The related segments consist mainly of Jews. As far as the segments regarding this cluster are concerned, the Jewish population's residences overlap with the Non-Residential Areas.

2. *Predominantly Armenian*

Armenians are overrepresented in the cluster groups in this section. There are 23 people in total in the C5 cluster, constituting 1.04% of the shoreline population (Table 4). The related segments are located in the vicinity of Arnavutköy on the western shoreline of the Bosphorus and in the vicinity of Karaköy on the northern shoreline of the Golden Horn. Although the related segments predominantly host Armenians, a certain ratio of Jews and Armenians live here together, as well.

There are 44 people in total in the C3 cluster, composing 1.99% of the population (Table 5). These segments are located in the vicinity of Ortaköy and Kireçburnu on the western shoreline of the Bosphorus and in the vicinity of Hasköy on the northern shoreline of the Golden Horn. Jews and Armenians reside together to a certain degree here, although the residents of the related segments are predominantly Armenians.

3. *Predominantly Armenian and Greek*

Armenians and Greeks are overrepresented and balanced in the cluster group in this section (Table 6). There are 66 people in total in the C10 cluster, which amount to 2.99% of the shoreline population. These segments are located in the vicinity of Emirgan (to the west of the Bosphorus), while also in the vicinity of Kadıköy (to the east of Bosphorus), and finally in Sütlüce (to the north of the Golden Horn). Armenian and Greek concentrations remain close to each other in these segments. In other words, Armenians and Greeks are represented in a balanced fashion.

Table 2. Jews 53.16% (4.53%)

TOTAL PEOPLE	%	CLUSTER		
79	3.58	C2		
ROW 3-2	**Jews**	**Non-Residential**	**Armenians**	**Muslims**
Concentration Index	20.31	-1.02	-1.14	-3.34
% Share in Ethno-Religious Group	42	2.92	1.2	1.36
% Share in the Cluster	53.16	31.65	1.27	13.92
% Share in the Shoreline Population	4.53	38.83	3.76	36.66
3114	Tekfurdağı İsk.-Tahtapuş İsk.			
3117	Balıkhane İsk.-Tüfenkhane İsk.			
3118	Tüfenkhane İsk.-Cibali İsk.			
3124	Balat İsk.-Kalafatçılar Meydanı			

Table 3. Jews 20.59% (4.53%) - Non-Residential Areas 52.94% (38.83%)

TOTAL PEOPLE	%	CLUSTER		
34	1.54	C12		
ROW 3-12	**Jews**	**Non-Residential**	**Greeks**	**Muslims**
Concentration Index	4.4	1.32	-0.91	-1.83
% Share in Ethno-Religious Group	7	2.1	0.91	0.74
% Share in the Cluster	20.59	52.94	8.82	17.65
% Share in the Shoreline Population	4.53	38.83	14.86	36.66
3116	Unkapanı İsk.-Balıkhane İsk.			
3205	Pîrî Mehmet Paşa İsk.-Hamam İsk.			

Table 4. Armenians 60.87% (3.76%) - Jews 8.7% (4.53%)

TOTAL PEOPLE	%	CLUSTER		
23	1.04	C5		
ROW 3-5	**Armenians**	**Jews**	**Greeks**	**Non-Residential**
Concentration Index	14.12	0.94	-0.23	-1.65
% Share in Ethno-Religious Group	16.87	2	0.91	0.47
% Share in the Cluster	60.87	8.7	13.04	17.39
% Share in the Shoreline Population	3.76	4.53	14.86	38.83
1016	Kuruçeşme İsk.-Bostani Ocağı			
3217	Mumhane İsk.-Eğrikapı İsk.			

Table 5. Armenians 29.55% (3.76%) - Jews 9.09% (4.53%)

TOTAL PEOPLE	%	CLUSTER			
44	1.99	C3			
ROW 3-3	Armenians	Jews	Muslims	Non-Residential	Greeks
Concentration Index	8.82	1.42	−0.28	−1.47	−2.17
% Share in Ethno-Religious Group	15.66	4	1.85	1.28	0.3
% Share in the Cluster	29.55	9.09	34.09	25	2.27
% Share in the Shoreline Population	3.76	4.53	36.66	38.83	14.86
1012	Ortaköy Camii-Koyun İsk.				
1040	Kireçburnu				
3206	Hamam İsk.-Hasköy İsk.				

Table 6. Armenians 7.58% (3.76%) - Greeks 19.7% (14.86%)

TOTAL PEOPLE	%	CLUSTER		
66	2.99	C10		
ROW 3-10	Armenians	Greeks	Non-Residential	Muslims
Concentration Index	1.6	1.02	0.86	−1.26
% Share in Ethno-Religious Group	6.02	3.96	3.5	2.22
% Share in the Cluster	7.58	19.7	45.45	27.27
% Share in the Shoreline Population	3.76	14.86	38.83	36.66
1027	Baltalimanı-Mirgünoğlu			
2041	Haydarpaşa-Kadıköy İsk.			
3201	Karaağaç Bahçesi-Sütlüce İsk.			
3203	Kumbaracı Camii-Buzhane İsk.			

4. *Predominantly Foreigner/Beratlı*

Foreigners/*Beratlı*s are overrepresented in the cluster group in this section (Table 7). There are 144 people in total in the C4 cluster, constituting 6.52% of the population. The related segments are located in the vicinity of Yeniköy and Büyükdere to the west of the Bosphorus and in the vicinity of Çengelköy on the east side of the Bosphorus. Although these segments are predominantly Foreigners/*Beratlı*s, they reside together with Greeks.

Table 7. Foreigners/*Beratlıs* 13.19% (0.95%) - Greeks 31.94% (14.86%)

TOTAL PEOPLE	%	CLUSTER
144	6.52	C4

ROW 3-4	Foreigners/ Beratlıs	Greeks	Boyars	Armenians	Jews	Non-Residential	Muslims
Concentration Index	15.06	5.32	0.54	0.25	−0.99	−2.13	−3.41
% Share in Ethno-Religious Group	90.48	14.02	11.11	7.23	4	4.67	3.46
% Share in the Cluster	13.19	31.94	0.69	4.17	2.78	27.78	19.44
% Share in the Shoreline Population	0.95	14.86	0.41	3.76	4.53	38.83	36.66

1013	Koyun İsk.-Defterdar Burnu
1030	İstinye İsk.-Mezarburnu İsk.
1033	Ekmekçi İsk.-Köy İsk.
1043	(Büyükdere_1) Büyükdere Çayırı-Mehmet Paşa Camii
1044	(Büyükdere_2) Mehmet Paşa Camii-Mezarburnu (Sarıyer Sınırı)
2026	Kulebahçe-Çengel Karyesi İsk.

5. *Predominantly Greek*

Greeks are overrepresented in the cluster group in this section (Table 8). There are 372 people in total in the C6 cluster, which means 16.86% of the shoreline population. The cluster covers the 13 segments written below the table, which are located to the west of the Bosphorus and south of the Golden Horn. In these segments Greeks and Boyars live together.

6. *Predominantly Muslim*

Muslims are overrepresented in the cluster group in this section (Table 9). There are 388 people in total in the C13 cluster, hosting 17.58% of the population. Spatially speaking, the cluster covers the 21 segments written below the table, which are located on the western and eastern shorelines of the Bosphorus, as well as on the southern and northern shorelines of the Golden Horn. These segments are predominantly Muslim.

There are 110 people in total in the C9 cluster, which corresponds to 4.98% of the population (Table 10). The related segments are located on the western and eastern shorelines of the Bosphorus, as well as on the northern shoreline of the Golden Horn. Although the Muslim population is higher in these segments in terms of ratio, as far as concentration is concerned, the Armenians and Muslims are close to each other. Armenians and Muslims dwelled together in these parts.

There are 423 people in total in the C11 cluster, that is, 19.17% of the shoreline population (Table 11). The 26 segments written below the table cover the western and eastern shorelines of the Bosphorus and the southern shoreline of the Golden Horn. In these segments, the Muslim population's residential areas overlap with the Non-Residential Areas.

Table 8. Greeks 68.01% (14.86%) - Boyars 2.15% (0.41%)

TOTAL PEOPLE	%	CLUSTER
372	16.86	C6

ROW 3-6	Greeks	Boyars	Armenians	Foreigner/ Beratlıs	Jews	Non-Residential	Muslims
Concentration Index	26.59	5.26	1.34	−0.82	−1.91	−6.53	−10.39
% Share in Ethno-Religious Group	77.13	88.89	22.89	9.52	9	7.7	1.85
% Share in the Cluster	68.01	2.15	5.11	0.54	2.42	17.74	4.03
% Share in the Shoreline Population	14.86	0.41	3.76	0.95	4.53	38.83	36.66

1018	Kuruçeşme Bostancı Ocağı-Arnavutköy İsk.
1019	Arnavutköy İsk.-Akıntıburnu
1034	Köy İsk.-Köybaşı (Yeniköy) İsk.
1035	Köybaşı (Yeniköy) İsk.-Kalender Bahçesi
1037	Tarabya Sınırı-Tarabya Mescidi
1038	Tarabya Mescidi-Tarabya İsk.
1039	Tarabya İsk.-Kireçburnu
1048	Yenimahalle
3119	Cibali İsk.-Ayakapısı İsk.
3120	Ayakapısı İsk.-Yenikapı İsk.
3121	Yenikapı İsk.-Petrokapısı İsk.
3122	Petrokapısı İsk.-Fenerkapısı İsk.
3123	Fenerkapısı İsk.-Balat İsk.

Table 9. Muslims 82.73% (36.66%)

TOTAL PEOPLE	%	CLUSTER		
388	17.58	C13		
ROW 3-13	**Muslims**	**Armenians**	**Non-Residential**	**Greeks**
Concentration Index	14.99	-3.56	-6.98	-7.46
% Share in Ethno-Religious Group	39.68	1.2	7.58	0.3
% Share in the Cluster	82.73	0.26	16.75	0.26
% Share in the Shoreline Population	36.66	3.76	38.83	14.86
1008	Çırağan Sarayı			
1010	Yahya Efendi Deresi-Ortaköy			
1022	Küçük Bebek-Kayalar			
1024	Rumelihisarı-Mahmut Han Camii			
1026	Şeytan Akıntısı-Baltalimanı			
1029	Tokmakburnu İstinye İsk.			
2001	Anadolukavağı			
2007	Hünkar İsk.-Yalıköy İsk.			
2008	Yalıköy İsk.-Beykoz İsk.			
2014	Çubuklu Bahçesi-Kanlıca İsk.			
2015	Kanlıca İsk.-Bahai Körfezi			
2016	Bahai Körfezi-Anadolu Hisarı			
2020	Göksu-Kandilli Camii			
2023	Vani Efendi Camii-Kulebahçe			
2028	Abdülhamid Camii-İstavros İsk.			
2029	İstavros İsk.-Nakkaş Paşa			
2032	Kuzguncuk-Öküz Limanı			
2037	(Üsküdar_4) Fatih Sultan Mehmet Han Camii-Harem İsk.			
3126	Ayvansaray İsk.-Yavedud İsk.			
3202	Sütlüce İsk.-Kumbaracı Camii			
3207	Hasköy İsk.-Fatih Sultan Mehmet Camii			

Table 10. Muslims 62.73% (36.66%) - Armenians 12.73% (3.76%)

TOTAL PEOPLE	%	CLUSTER			
110	4.98	C9			
ROW 3-9	**Armenians**	**Muslims**	**Jews**	**Greeks**	**Non-Residential**
Concentration Index	4.85	4.52	−1.34	−2.81	−3.48
% Share in Ethno-Religious Group	16.87	8.53	2	1.52	2.33
% Share in the Cluster	12.73	62.73	1.82	4.55	18.18
% Share in the Shoreline Population	3.76	36.66	4.53	14.86	38.83
1015	Ekmekçioğlu Deresi-Kuruçeşme İsk.				
1031	Mezarburnu İsk.-Pazarkayığı İsk.				
2022	Şattiler (Akıntıburnu)-Vani Efendi Camii				
2027	Çengel Karyesi İsk.-Abdülhamid Camii				
2042	Kadıköy İsk.-Moda Burnu				
3204	Buzhane İsk.-Pîrî Mehmet Paşa İsk.				

Table 11. Muslims 58.87% (36.66%) - Non-Residential Areas 40.66% (38.83%)

TOTAL PEOPLE	%	CLUSTER		
423	19.17	C11		
ROW 3-11	**Muslims**	**Non-Residential**	**Jews**	**Greeks**
Concentration Index	7.54	0.6	−4.15	−7.8
% Share in Ethno-Religious Group	30.78	20.07	1	0.3
% Share in the Cluster	58.87	40.66	0.24	0.24
% Share in the Shoreline Population	36.66	38.83	4.53	14.86

1001	Tophane-Salıpazarı
1002	Salıpazarı-Fındıklı
1003	Fındıklı-Kabataş
1004	Kabataş-Karabali
1006	Hayrettin İsk.-Beşiktaş İsk.
1007	Beşiktaş İsk.-Çırağan Sarayı
1009	Çırağan-Yahya Efendi Deresi
1014	Defterdar Burnu-Ekmekçioğlu Deresi
1020	Akıntıburnu-Bebek
1021	Bebek-Küçük Bebek
1025	Rumelihisarı-Şeytan Akıntısı
1046	Mezarburnu-Sarıyer İsk.
1047	Sarıyer İsk.-Yenimahalle
1050	Rumelikavağı
2009	Beykoz İsk.-Sultaniye
2011	İncirköy İsk.-Paşabahçe İsk.
2017	Anadoluhisarı
2019	Göksu
2021	Kandilli Camii-Şattiler (Akıntıburnu)
2034	(Üsküdar_1) Öküz Limanı-Mihrimah Sultan Camii
2035	(Üsküdar_2) Mihrimah Sultan Camii-Şemsi Paşa Camii
2036	(Üsküdar_3) Şemsi Paşa Camii-Fatih Sultan Mehmet Han Camii
3101	Yalı Kasrı-Ocaklar
3115	Tahtapuş İsk.-Unkapanı İsk.
3127	Defterdar İsk.-Eyüp İsk.
3128	Eyüp İsk.-Çukur Saray

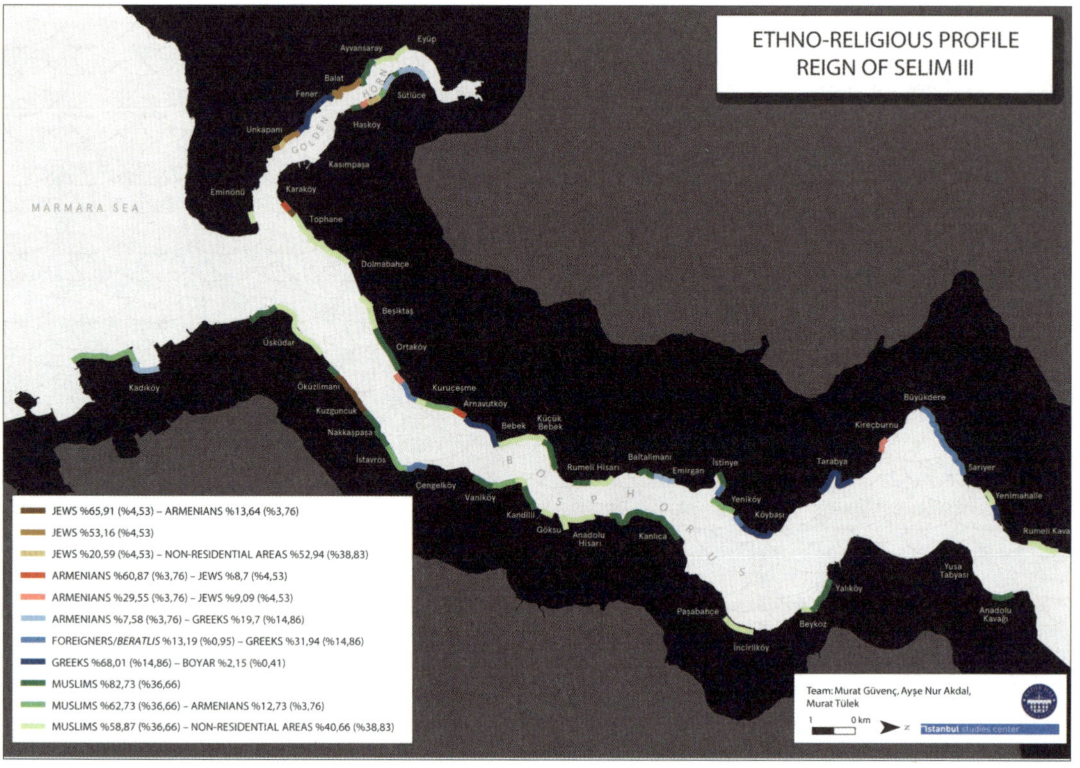

Fig. 3. The shorelines of Istanbul during the reign of Selim III (1789–1808) according to the *bostancıbaşı* register.

Conclusion

When the tables are assessed, it is obvious that ethnic-religious elements are concentrated in different regions of Istanbul's shoreline to different degrees (Fig. 3). There are segments distinguished by the overrepresentation of a single ethnic-religious group, as well as segments marked by different ethnic-religious groups where more than one group is overrepresented at the same time to different degrees, such as Armenians and Jews, Armenians and Greeks, or Armenians and Muslims. An interesting exception to this is the absence of any segments where Jews and Greeks are overrepresented simultaneously. The prevalence of segments where Muslims are overrepresented on the map demonstrates the effect of "Islamization" of the shorelines during the seventeenth and eighteenth centuries, as discussed above.

Introducing an alternative perspective on Ottoman historiography, our research shows how digital data from *bostancıbaşı* registers can be reevaluated when processed through methods employed in digital humanities. The method we have used not only makes socio-spatial differentiation on the shoreline ready for mapping but also contributes significantly to the socio-spatial review of Istanbul's shorelines in the pre-modern era. Hence, our research will be a reference for those studying urban history and, especially, the shorelines of Istanbul.

Having said that, it should be kept in mind that research findings reported here are confined to narrow strips of the shoreline. Future research that could integrate a study of the inner parts of the city would shed further light on the subject matter. The registers are shaped from the viewpoints of those who kept them, namely the *bostancıbaşıs* and their clerks. No matter how internally systematic a register's structure seems to be, it is not the result of modern census methods. For instance, while there are abundant details regarding certain issues, others are just overlooked, with no measures or quantitative information in the modern sense. What is more, the diverse nature of the data presented makes classification even harder, forcing the researcher to choose between focusing on the detail or the readability of the resultant maps. Another challenge is how the classification of the data at hand depends on a researcher's initiative regarding certain aspects. A historian needs to decide on the best ways to abstract the information under various categories. Needless to say, striking the perfect balance between loyalty to the original document and abstraction has been the toughest challenge presented by the research. In conclusion, the exploratory aspect of our research cannot be overstated. One must not forget that, as far as new mapping approaches are concerned, there is never a single "perfect" map, and that the map produced here is regarded as one of many that could exist.

Acknowledgments

We would like to express our sincere gratitude to Prof. Çiğdem Kafescioğlu, Prof. Edhem Eldem, and Prof. Cengiz Kırlı for their valuable comments. We would like to extend our warmest thanks to Murat Tülek, Ayşe Tuğba Silahtar, Kalliopi Amygdalou, Mehmet Sağçolak, Fırat Kaplan, and Matt Hanson for their valuable contributions.

Works Cited

Artan, Tülay. 2010. "18. yy'da İstanbul: Uzlaşma ve Yeniden Yapılanma Dönemi." In *Bizantion'dan İstanbul'a Bir Başkentin 8000 Yılı*, edited by Şevki Koray Durak, 300-13. Istanbul: Sakıp Sabancı Museum.

Aydın, Yusuf Alperen. 2016. *Müteşebbis ve Devlet: 18. Yüzyılda Peksimetçibaşı Nureddin Ağa'nın Yükselişi*. Istanbul: Kitabevi.

Bardakçı, Murat. 2013. Üçüncü *Selim Devrine Ait Bir Bostancıbaşı Defteri*. Istanbul: Pan Press.

Başaran, Neriman. 1951. "Defter-i Bostancı: Anadolu ve Rumeli Sevahilinde Bulunan Binalara Dair Defter." BA thesis, Istanbul University.

Eldem, Edhem. 1999. "18. Yüzyıl ve Değişim," *Cogito* 19 (Summer): 189-99.

———. 1996. "Osmanlı Dönemi İstanbulu." In İstanbul *Armağanı 3: Gündelik Hayatın Renkleri*, edited by Mustafa Armağan. Istanbul: İstanbul Büyükşehir Belediyesi.

Hamadeh, Shirine. 2007. *Şehr-i Sefâ*. Istanbul: İletişim Yayınları.

Hovannesyan, Sarkis Sarraf. 1996. *Payitaht İstanbul'un Tarihçesi*. Istanbul: Tarih Vakfı Yayınları.

İnciciyan, Gugas. 2000. *Boğaziçi Sayfiyeleri*. Istanbul: Eren Yayıncılık.

———. 1976. *18. Asırda İstanbul*. Istanbul: Baha.

Kafescioğlu, Çiğdem. 2011. "Osmanlı Şehir Tahayyülünün Görsel ve Edebi İzleri: Onaltıncı ve Onyedince Yüzyıl Menzilname ve Seyahatnamelerinde Şehir İmgeleri." In *Kültürel Kesişmeler ve Sanat*, edited by Serpil Bağcı and Zeynep Yasa Yaman, 139-50. Ankara: Hacettepe University Press.

Kayra, Cahit and Erol Üyepazarcı. 1992. *II. Mahmut'un İstanbul'u Bostancıbaşı Sicilleri*. Istanbul: İBKY.

Kırlı, Cengiz and Betül Başaran. 2011. "18. Yüzyıl Sonlarında Osmanlı Esnafı." In *Osmanlı'dan Cumhuriyet'e Esnaf ve Ticaret*, edited by Fatmagül Demirel, 7-20. İstanbul: Tarih Vakfı Yurt Yayınları.

Koçu, Reşat Ekrem. 1958a. "Bostancıbaşı Defterleri." İstanbul *Enstitüsü Dergisi* 4: 39-90.

———. 1958b. *İstanbul Ansiklopedisi*. Istanbul: Tan Matbaası.

Kömürciyan, Eremya Çelebi. 1988. *İstanbul Tarihi: 17. Asırda İstanbul*. Istanbul: Eren.

Lebart, Ludovic. 1994. "Complementary Use of Correspondence and Cluster Analysis." In *Correspondence Analysis in the Social Sciences: Recent Developments and Applications*, edited by Michael Greeenacre and Jvrg Blasius, 162-78. London: Academic Press.

Orbay, İffet. 2011. "İstanbul'a Bakış: Kitab-ı Bahriye Yazmalarında İstanbul Tasvirinin Ortaya Çıkışı ve Evrimi." In *Gelenek, Kimlik, Birleşim: Kültürel Kesişmeler ve Sanat - Günsel Renda'ya Armağan*, edited by Zeynep Yasa Yaman and Serpil Bağcı, 199-208. Ankara: Hacettepe University Press.

Rado, Şevket. 1972. "Bostancıbaşı Defteri: 1802 Yılında Boğaziçi ve Haliç Kıyılarında Kimler Oturdu." *Hayat Tarih Mecmuası* 6: 3–32.

Sertel, Vuslat. 1951. "Defter-i Bostancı: Anadolu ve Rumeli Sevahilinde Bulunan Binalara Dair Defter." BA thesis, Istanbul University.

Süer, Hale. 1951. "1206 Senesinde Karaağaç'tan Rumeli Kavağına Kadar Sahilhaneler Esamesi." BA thesis, Istanbul University.

Yerasimos, Stefanos. 2010. *İstanbul İmparatorluk Başkenti*. Istanbul: Tarih Vakfı.

IV

Mapping Cultural Heritage and Identity

Mapping the Jewish Communities of Medieval Anatolia

Nicholas de Lange

Introducing the Project

The project "Mapping the Jewish Communities of the Byzantine Empire," which was based in the University of Cambridge under my direction, ran from 2010 to 2013, and was financed by a generous grant from the European Research Council. This mapping project covered the entire territory of the Byzantine Empire. In the remarks that follow I shall focus particularly on the section of the project dealing with Anatolia.

The project began from the perception that the history of the Byzantine Jewish communities suffered from many inadequacies. There are various reasons for this (de Lange 1989). Although Byzantium was a central and very important part of the medieval Jewish world, very little has been written about this subject. Moreover, the few existing studies tend to suffer from a lack of rigor and accuracy. To give just one example, one of the few books on the subject states that "for most of our period" (i.e., from the early sixth to the early thirteenth century) the important communities were "at Amorium, Attaleia, Seleucia and smaller centres of central and southern Asia Minor" (Sharf 1971, 2). In fact our research showed, as we shall see later, that the community in Amorium (or "Amorion") is only attested in the early ninth century, the community in Attaleia is not attested before 1020, and the community in Seleucia only during the tenth and eleventh centuries. That does not mean, of course, that these communities did not exist at other times, too, because the historical record is incomplete; but there is no sound basis for making the claim put forward in the book. This kind of unreliable generalization unfortunately pervades the relevant literature.

It was also observed that the maps accompanying historical narratives were often vague or inaccurate in the way they located places. They were simple sketch maps in black and white, very limited by this very fact in the information they were able to convey, and generally showing only the coastline and some place-names. Other physical features were not marked, and few regions or administrative divisions were named.

Consequently, we planned to assemble the totality of evidence for the existence of all the Byzantine Jewish communities at a particular time and place, and then to use the data to construct reliable maps, based on satellite imagery, and using Geographic Information Systems (GIS). The resulting maps would be made freely available via the internet, together with the evidence on which they are based.

The project thus makes a contribution to digital history, a branch of digital humanities. In addition to the inherent interest of the subject, it can be seen as a pilot project applying GIS to a historical topic, along the lines of earlier work (e.g., Gregory and Ell 2007). The subject chosen is attractive for this purpose because it has a finite and very limited body of evidence. From the outset the project was conceived in these terms. We hope that the methods used can be applied to other historical projects. We also hope that our maps can in the future interact or interlace with other historical maps conceived in the same or similar terms.

Using GIS

Our project used GIS to highlight the temporal and spatial dimensions of the data. Our GIS stores two types of information: *features* (broadly speaking these are references to events), which have a specific location, and *attributes*, which are data attached to these locations (such as a building, institution, name, or occupation). Although desktop GIS has been applied in many areas of academic research, the technology has largely remained the preserve of specialists, as it can be difficult to use, requiring training and experience. The application of GIS to humanities research in general, and to the interpretation of historical data in particular, has the potential to develop and answer innovative research questions.

The "Mapping the Jewish Communities of the Byzantine Empire" project data are available through web maps at byzantinejewry.net. The resulting web maps employ a sophisticated search and browsing mechanism developed using JavaScript in collaboration with Humlab, based at Umeå University in Sweden. The web maps are *dynamic* and *interactive*. Time and space primarily structure how the data are displayed. A time bar allows users to limit their search to a particular period. It is also possible to limit the search to one or more places, with provision to select alternative toponyms, such as ancient or modern names. Other filters permit users to narrow their search further and tailor it to their specific needs, for example by specifying occupations or types of residence. In this way we hope to make our simple maps into a potentially very useful research tool (Fig. 1).

The filters create requests that query our relational PostgreSQL database. Modeling spatial and temporal uncertainty required a relatively complex structure for the size of the dataset, leading the database to include more than twenty tables. The database uses the PostGIS plugin for PostgreSQL which allows our MapServer (mapserver.org) to provide a Web Map Service to display points on the Google Maps API component in the browser. If preferred, the data can be accessed in tabular form, and further information is supplied in the associated data tables, such as names of individuals mentioned in the sources and references to published editions.

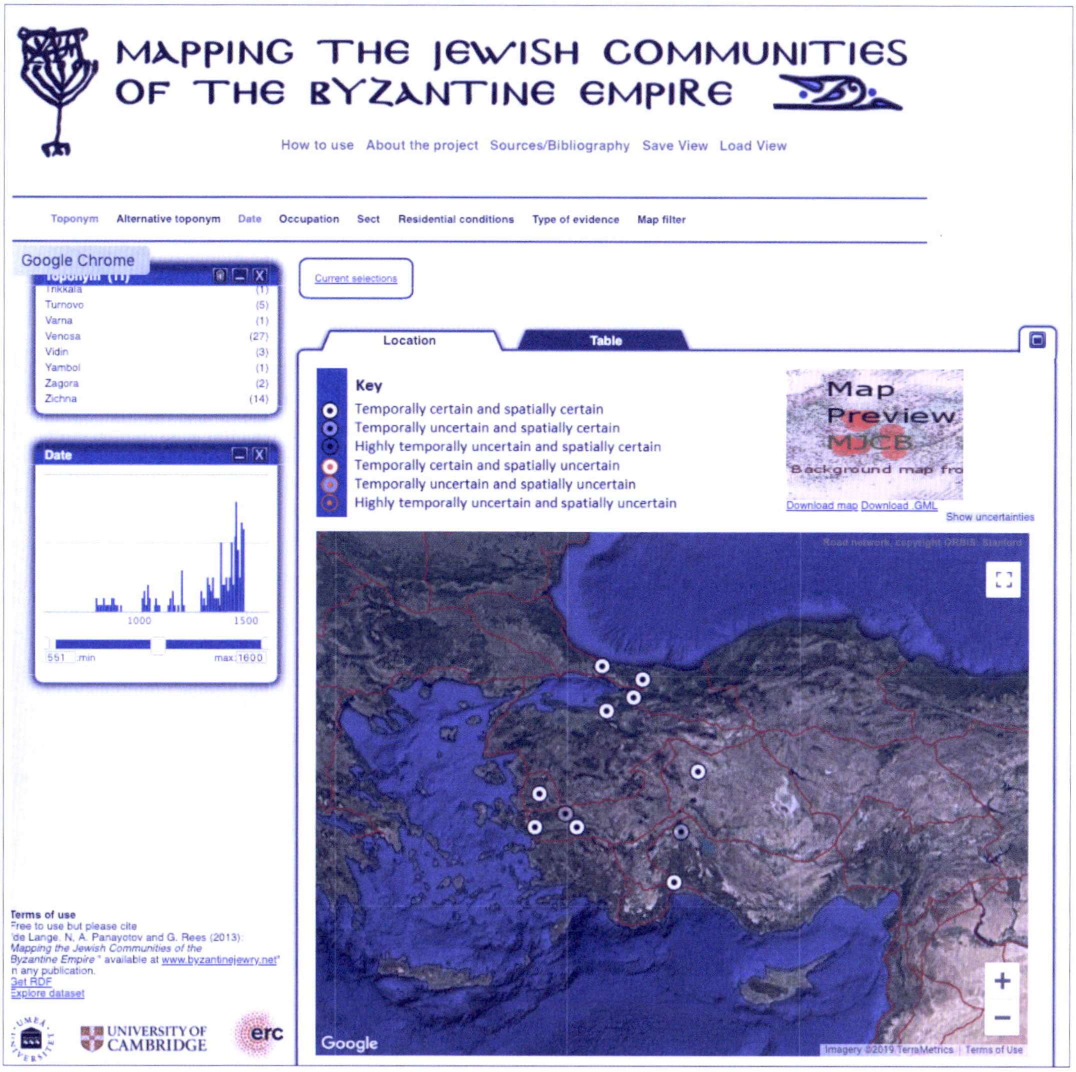

Fig. 1. The mapping portal (screenshot).

Maps and tables are downloadable through the system and the underlying data have been made available on the digital data repository, Zenodo (Rees et al. 2017). The database has also been made available in the RDF linked data format and linked to the Pleiades gazetteer of the ancient world (Pleiades, n.d.). Linking these data can facilitate integration with broader historical evidence and environmental or archaeological data. In this way our pilot project can make a contribution to other, perhaps wider, projects concerned with the geographical areas in question.

Selection of Data

Our explicit aim was to assemble the totality of evidence for these Jewish communities, yet this statement is subject to one important qualification, already implied. It was a fundamental principle that all features selected for inclusion had both spatial and temporal details. We were obliged to exclude on principle all data that had no place or date attached to them. It was essential for us to employ a rigorous policy of exclusion of data that could not be both dated and localized so that our maps would be accurate and reliable. A historical narrative can afford a certain vagueness; a map cannot. Published histories of Byzantine Jewry have a notable tendency to rely on generalization, extrapolation, guesswork and, occasionally, pure fantasy, simply because the data are so limited. By enforcing a rigorous selection of data, we hoped to make a serious contribution to clearing away the fog and establishing the history of the Jewish presence in Byzantium on a sound basis.

Data Uncertainties

Two types of data uncertainty called for special treatment: spatial and temporal. In the first place, some features were defined by a limited amount of uncertainty as to their location. This uncertainty took various forms:

1. The source could refer to a place-name that is ambiguous. For example, a Hebrew source dated 1362 refers to the city of Laodikeia. A reference work we used comments: "There are two cities named Laodikeia in Anatolia: one is just north of Konya (Cecaumene); the other is on the Maeander River, southwest of Mastaura and south of Philadelphia" (Bowman 1985, 285n11). The author of the work did not attempt to identify the Laodikeia in question. We therefore had two options: either exclude this information on the grounds that it was not precise enough to locate it on the map or find a way to include it on the map without misleading users. We opted for the second alternative.

But how should we proceed? We believed that it was not up to us to decide which of the two possible locations was correct. It should be stressed that, generally speaking, we carried out no original research for this project, but only represented what we found in reliable editions and reference works. In all probability, there is no way of deciding which of these two cities called Laodikeia had a Jewish community in the mid-fourteenth century. On the other hand, if we put both cities on the map, that would suggest that both cities had a Jewish community, which was not necessarily the case. We pondered long and hard over this problem and came up with a solution: we would put both cities on the map but identify them with a special symbol which would point to the spatial uncertainty and thus avoid potentially misleading users. Once we had decided on this solution, we could apply it to other types of spatial uncertainty as well (Fig. 2).

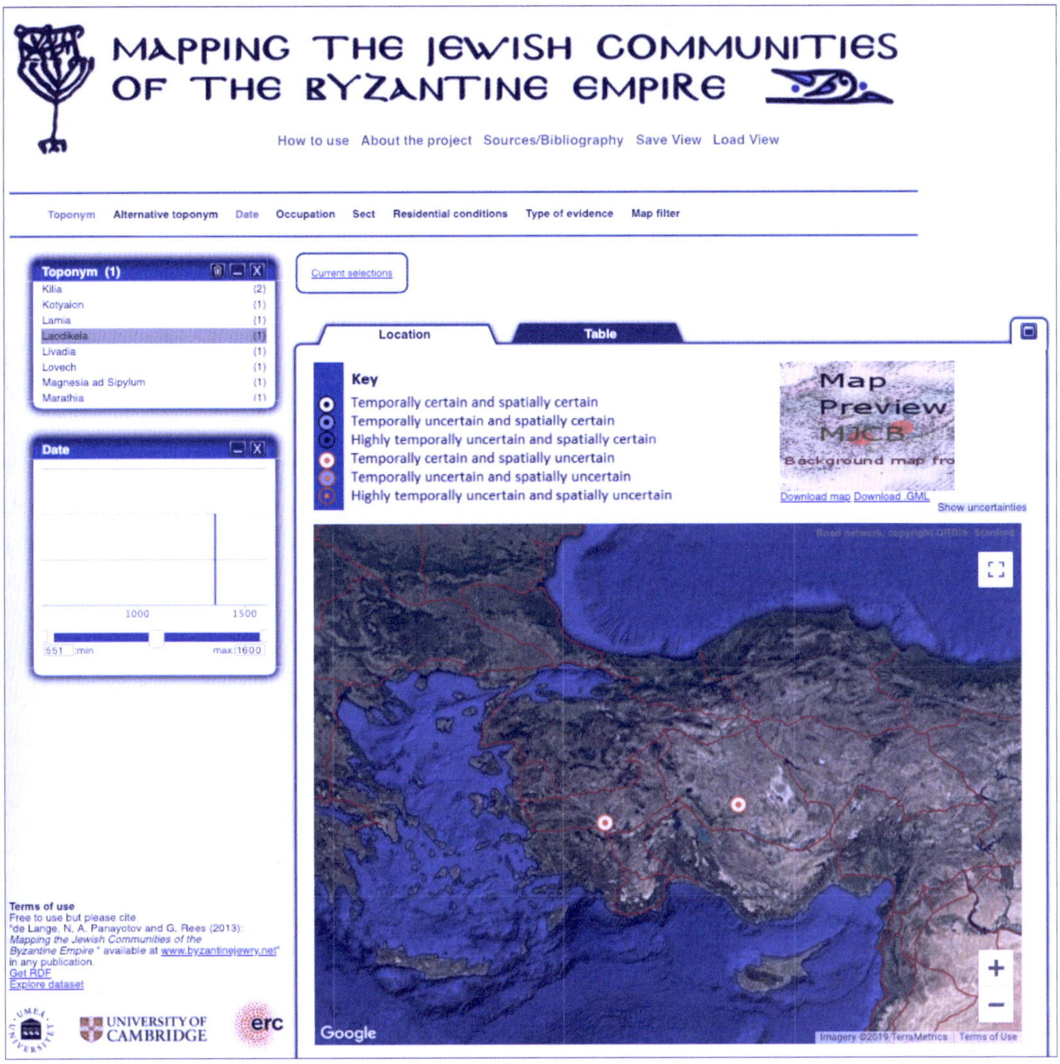

Fig. 2. Map showing a place name with spatial uncertainty: Laodikeia.

2. A Hebrew manuscript bears a colophon stating that it was copied in 1207 "in the city of Gagra." No city of this name is known. One reference work identified it as Gagry, on the eastern shore of the Black Sea. Another, however, identified it as Gangra (Germanicopolis) in Paphlagonia, near the southern shore of the same sea. Once again, it was not for us to determine which of the two secondary sources was right: we put both locations on the map, using the same distinctive symbol for spatial uncertainty.

3. Another type of case is where only one location had been proposed by modern authorities, but without complete certainty. An example is a letter dated 21 July 1137 in Seleukeia in Cilicia (Silifke), mentioning a place located some 120 km away from that

town. The editor read the name of the place as "Palaia," and we interpreted the reference as being to the city of Isaura Palaia (Leontopolis), identified as the modern site of Zengibar Kalesi near Konya. Because there is some uncertainty about the identification of the place mentioned in the letter, however, we used the same symbol for spatial uncertainty (Fig. 3).

Because in the Jewish reckoning the year begins in September, whereas we use the current Christian (CE) system beginning in January, temporal uncertainty was a pervasive problem as well. Consequently, if a source provided no date in the year, we could not be certain whether, for example, an event mentioned as happening in the year 5000 in a Hebrew document occurred in the year 1239 or 1240 CE. Clearly the map should not display both years, as the event only occurred in one of them. And so we resorted to a similar solution to that used for spatial uncertainty: we used a distinctive symbol, of a different color, to indicate temporal uncertainty.

This symbol was useful in other circumstances too. It happened not infrequently that an event mentioned in a source could not be pinned down to a single year but was placed by historians within a range of years, for example in the reign of a particular emperor. In such cases, too, we were now able to include the event on the map, while indicating the temporal uncertainty.

While we were working on these cases of spatial and temporal uncertainty, we came across some events that were both spatially and temporally uncertain. Once we had established the principle of using different symbols, it was not difficult to expand the range of symbols. In the end we used six different symbols to indicate the various possible combinations and also degrees of uncertainty:

- Temporally certain and spatially certain [the most common symbol]
- Temporally uncertain and spatially certain
- Highly temporally uncertain and spatially certain
- Temporally certain and spatially uncertain
- Temporally uncertain and spatially uncertain
- Highly temporally uncertain and spatially uncertain

In this way we were able to exploit as many of the data as possible while maintaining a high level of accuracy and reliability. In each and every case the user is able to access references to the data sources and publications we used. By this means we tried to make the maps a useful tool for specialized research while also making them accessible to a wide audience.

Sources

In assembling the data we first, naturally, had to determine the sources we would use. We cast our net as widely as possible so as to include any events that had a date and a

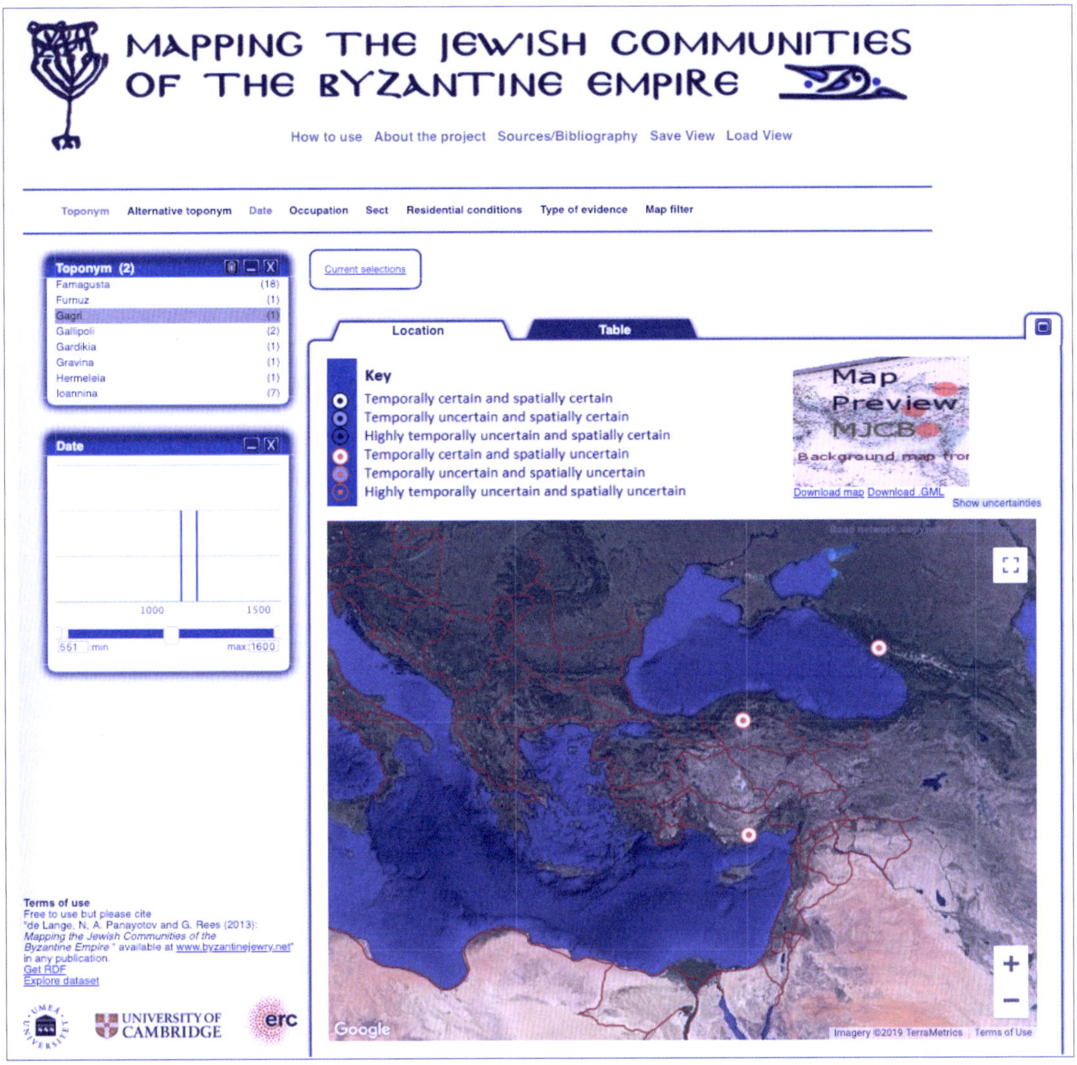

Fig. 3. Map showing place names with spatial uncertainty: Gagra and Palaia.

place attached to them. The sources included both Christian and Jewish literary sources and documents, as well as other types of sources such as inscriptions and colophons in manuscripts (Fig. 4).

By way of illustration, we have two items relating to Amorion (or "Amorium," modern Hisar, near Emirdağ) in Phrygia. One comes from an anonymous Greek Christian historical work, known as "Theophanes Continuatus," devoted to events in the ninth century (Starr 1939, 98–99, no. 20). It mentions that there was a large Jewish community in Amorion, which maintained close relations with a local Christian sect, the Athinganoi. The only other reference to Jews in Amorion describes how they clashed

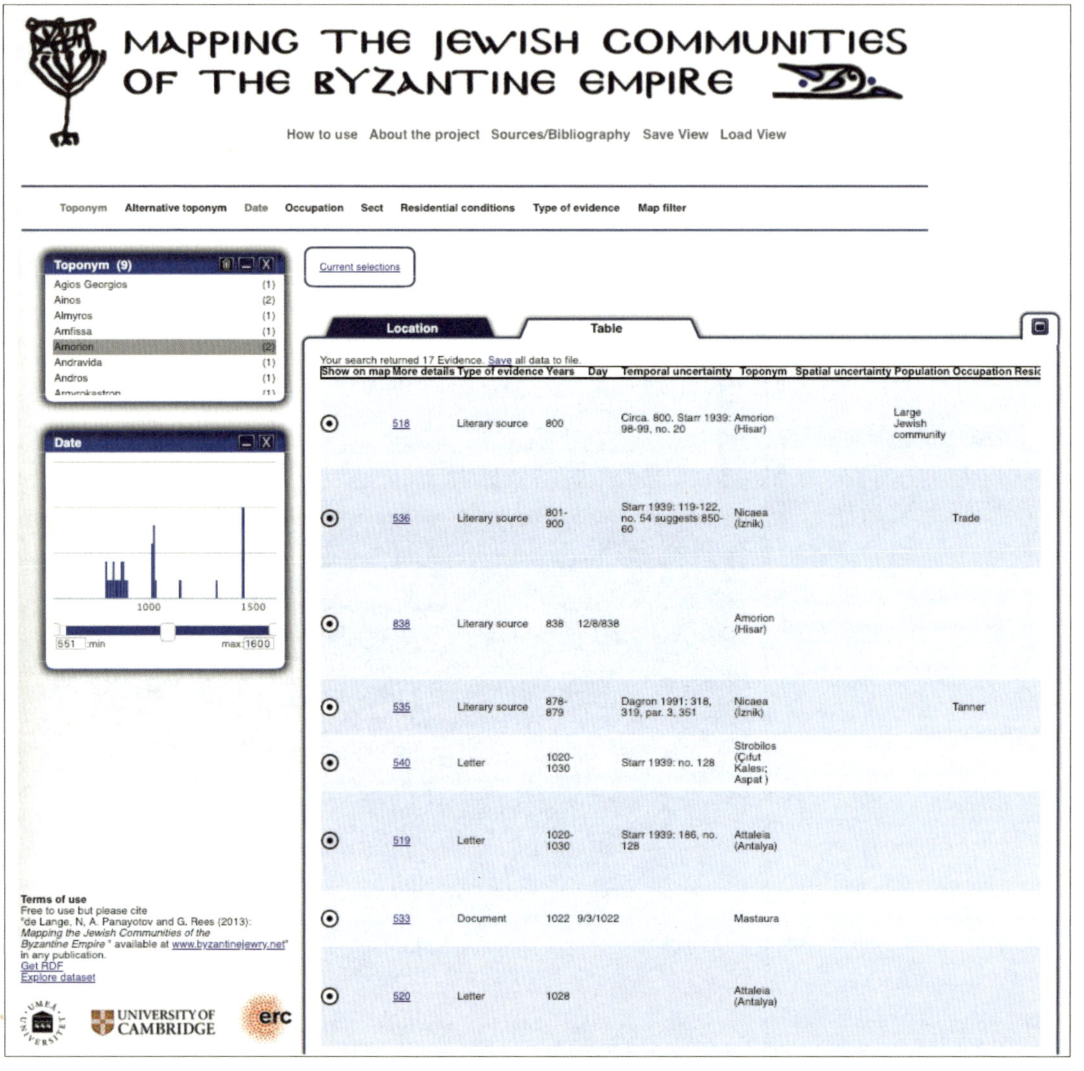

Fig. 4. Tabular data view showing sources.

with Christians on the eve of the conquest and destruction of the city by the Abbasid caliphate in 838 (Starr 1939, 109, no. 38). The source is another Greek Christian author, known as Genesios. What happened to this once large Jewish community after the Arab conquest we have no idea. As remarked earlier, there are no grounds for claiming that it existed until the thirteenth century.

A second example relates to Nicaea (modern İznik). It comes from yet another anonymous Greek Christian source, the life of St. Constantine "the Jew" (Starr 1939, 119–22, no. 54). It mentions that "some Hebrews lived there for the sake of its trade and other advantages." The same source refers to Synnada (Şuhut), where the saint was born in the Jewish community. (No other source refers to Jews in this Phrygian town.) The only other datable reference to Jews in Nicaea comes from the Greek Christian treatise of Gregory of Nicaea on the baptism of the Jews and is in fact the earliest source connecting Jews in Byzantium with the despised craft of tanning leather, with which they were later commonly associated (Dagron 1991, 318-19, 351). Both these sources relate to the ninth century. As with Amorion, we do not know what happened to this Jewish community after that.

That, then, is one kind of source: Christian writings in Greek. Completely different from these are Jewish documents, usually written in Hebrew or Arabic. Prominent among these are the documents recovered from the Genizah (a kind of rubbish dump) in a synagogue in Old Cairo. The Cairo Genizah documents have transformed our knowledge of Jewish life in the eleventh and twelfth centuries. Only a few of them relate to Byzantium, but they are of inestimable importance.

For example, we have some letters from the Genizah concerning Jews who had been captured by Arab pirates and offered for ransom. Two of these, datable to the 1020s, refer to Jews originating from Strobilos (Aspat or Çıfıt Kalesi, near Bodrum) and Attaleia (Antalya) (Starr 1939, 186, no. 128). These are the earliest references to a Jewish presence in both these places.

The next item is a marriage document, written in Mastaura near the river Maeander in 1022. It is the only source mentioning Jews in this place, and so its testimony is very precious (see Prinzig 1997, esp. 158). This kind of stray evidence is a reminder of how ignorant we are of many details in the story of Jewish settlement in Anatolia in the Middle Ages.

This fact is particularly powerfully brought home by an unexpected source: the Ottoman poll-tax registers (*cizye tahrir defterleri*) from Istanbul. After the conquest, Sultan Mehmed the Conqueror repopulated the depleted city by a policy of forcible transfer (*sürgün*) of populations from different parts of his empire, including the Jewish communities of Anatolia. In fact this policy, carried out over a period of some twenty years after the conquest, more or less spelled the end of the Byzantine Jewish communities scattered around Anatolia because, once moved to the capital, they were generally not allowed to retain ties with or to return to the towns where they had previously lived. In Istanbul, however, they retained an identity based on their previous residence. The tax registers recognize this identity and list the contributions by communities. Even well into the seventeenth century they list the Jews by congregations which carry the names of the towns from which they came, such as Tire, Sinop, Antalya, or Bursa. Study of these registers reveals the names of former communities for which there is no other evidence. An example is Kastamonu in Paphlagonia: the only trace of this Byzantine Jewish community is from an early tax register, after it had been forcibly transferred to Istanbul. The same is true of Furnuz, a place of uncertain location apparently close to Süleymanlı. Other members of this group are Borlu (Manisa Province) and Eğirdir

(Akrotiri, in Isparta Province). In all these cases we would have no evidence of a Jewish community were it not for their appearance in the Istanbul tax registers. By the time they make their appearance in these registers, the communities had ceased to exist in their original locations.

Chronological Framework

Mention of the *cizye* registers leads to one final point, the chronological framework of our maps. The starting point, so far as Anatolia is concerned, is effectively 800, since we have no earlier data from the medieval period. (A few locations in Byzantine Italy have data in the seventh and eighth centuries.) The end date of the project, however, raised some issues, and these relate to the question of the geographical limits of the project. Although the title of the project mentions the Byzantine Empire, it would have been impossible for us to restrict it to the Byzantine Empire as a political entity. The frontiers of the empire were constantly being fought over, and they changed dramatically with time. Towns were lost to the empire and subsequently regained. While a historical narrative can limit itself to the areas under Byzantine rule, the maps cannot. So when should the period covered by our maps end? The date conventionally accepted for the end of the Byzantine Empire, the conquest of Constantinople in 1453, has no relevance at all for Anatolia, where, for example, Bursa became the Ottoman capital in 1326. In any case, the culture and identity of the Greek-speaking Byzantine Jews, the so-called Romaniots, did not change from one day to the next with Seljuk or Ottoman conquest. The really radical change for the Jews came with the immigration of large numbers of western Jews from Spain and Portugal in the last decade of the fifteenth century. Consequently we decided that our maps should end at this point, regardless of who was ruling territory in various places at earlier times. And so the maps cover the period to the end of the fifteenth century in all the territories that belonged at one time or another to the Byzantine Empire, with the exception of a few relatively insignificant territories (in Jewish terms), such as Crimea, which raised too many problems of a historical and a cartographic nature. This choice gives a coherence to the project, both chronologically and spatially.

Conclusions

To sum up, our project aimed to make a significant contribution both to Byzantine and to Jewish history, by making available for the first time reliable, evidence-based maps of the Jewish presence in the constituent territories of the Byzantine Empire, both in Asia and in Europe. The project also aimed to make a contribution to digital humanities, and more specifically to digital history. Its major aim was to offer interactive maps, freely accessible online, and facilitating tailor-made searches by specialists and by the general public, using an advanced search and browsing mechanism. Through a comprehensive and rigorous selection of data it was designed as a reliable tool on which further research could be built, by scholars and others. In this regard a noteworthy feature was its pioneering exploitation of techniques for dealing with spatially and temporally *uncertain data*, so as to exploit a wide range of data without sacrificing accuracy and intelligibility. As well as being historically and geographically more reliable than existing maps, its usefulness to researchers was enhanced by the incorporation of references to all the relevant documentation and to key additional elements derived from the sources, such as aspects of the urban landscape, occupations, and even names of individuals (Fig. 5). Finally, through the use of *linked data* it was designed to interact with other, analogous projects, so that potentially it can function as part of a mosaic of digital historical maps.

Having been funded for a limited period, the maps are now a closed resource, because the team that created them has been disbanded and there is no provision for updating them. This is a regrettable feature of many projects funded in this way. It is to be hoped that in the future means can be developed to maintain such online resources, keeping them up to date and ensuring that they remain abreast of technical advances.

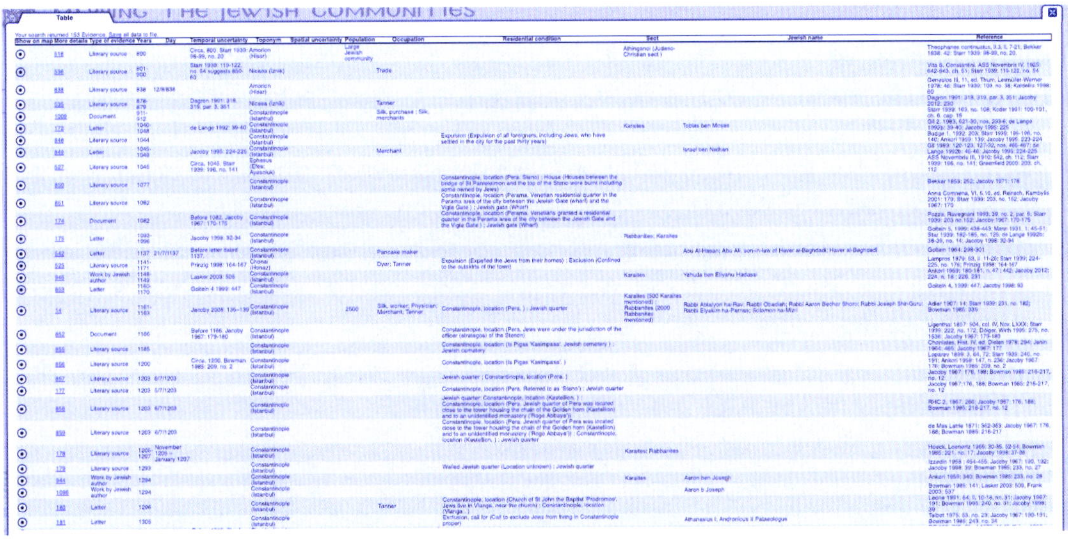

Fig. 5. Tabular data view showing various types of information.

Acknowledgments

I should like to acknowledge with warm thanks the contributions of my research associates Gethin Rees and Alexander Panayotov, as well as those of İnci Türkoğlu, our local researcher for Turkey. Their work underlies much of this chapter. More information about the project herein described can be found on the project website: mjcb.eu.

Works Cited

Bowman, Steven B. 1985. *The Jews of Byzantium 1204-1453*. Tuscaloosa: Alabama University Press.

Dagron, Gilbert, 1991. "Le traité de Grégoire de Nicée sur le baptême des Juifs." *Travaux et Mémoires* 11: 313-57.

de Lange, Nicholas, Alexander Panayotov, and Gethin Rees. 2010. Mapping the Jewish Communities of the Byzantine Empire (website). www.mjcb.eu.

———. 2013. Mapping the Jewish Communities of the Byzantine Empire (database). www.byzantinejewry.net.

de Lange, Nicholas. 1989. "Qui a tué les Juifs de Byzance?" In *Politique et Religion dans le Judaïsme antique et médiéval*, edited by Daniel Tollet, 327-33. Paris: Desclée.

Gregory, Ian N., and Paul S. Ell. 2007. *Historical GIS: Techniques, Methodologies and Scholarship*. Cambridge: Cambridge University Press.

Open Source Geospatial Foundation. 2019. MapServer. https://mapserver.org/.

Pleiades. n.d. Accessed February 26, 2015. https://pleiades.stoa.org/.

Prinzig, Günter. 1997. "Zu den Minderheiten in der Mäander-Region während der Übergangsepoche von der byzantinischen zur seldschukisch-türkischen Herrschaft (11 Jh.-Anfang 14. Jh)." In *Ethnische und religiöse Minderheiten in Kleinasien von der hellenistischen Antike bis in das byzantinische Mittelalter*, edited by Peter Herz and Jörn Kobes, 153-77. Wiesbaden: Harrassowitz.

Rees Gethin, Nicholas de Lange, Alexander Panayotov, and Fredrik Palm. 2017. "Mapping Jewish Communities of the Byzantine Empire" [Dataset]. Zenodo. http://doi.org/10.5281/zenodo.1116960.

Sharf, Andrew. 1971. *Byzantine Jewry: From Justinian to the Fourth Crusade*. London: Routledge & Kegan Paul.

Starr, Joshua. 1939. *The Jews in the Byzantine Empire, 641-1204*. Athens: Verlag der "Byzantinisch-neugriechischen Jahrbücher."

The Turkey Cultural Heritage Map of the Hrant Dink Foundation

Vahakn Keshishian

Introduction

The Turkey Cultural Heritage Map is a product of the Hrant Dink Foundation's cultural heritage inventory project. It can be accessed online at turkeyculturalheritage.com. The map is an online interactive research tool for both scholars and enthusiasts. The online map records Armenian, Greek, Syriac, and Jewish cultural heritage to make visible the multiple cultures and facets of different communities in Anatolia. The online map was developed as part of two projects. The first of these was called "Revealing and Advocating the Multicultural Heritage of Anatolia;" it ran between 2014 and 2016 and was funded by the European Union Ministry of the Republic of Turkey, under the scheme of Civil Society Dialogue, with additional funding provided by the Chrest Foundation. The follow-up project was called "Creative Reuse of Sites of Memory;" it ran between 2016 and 2018 and was funded by the European Union, with additional funding again provided by the Chrest Foundation.

On the interactive online map, users can search through an inventory, the results of which are visualized with their geolocations (Fig. 1). Other than the basic search, a variety of filtering criteria can be used to browse the map: by particular communities, a building's original function(s), or province, for instance. The result is that by using different combinations of search and browse functions, different maps can be visualized.

The online map project began with a phase of literature review, during which several lists and databases were identified to form the main sources of the inventory; these are detailed below. These lists were researched and compiled in the inventory with approximate coordinates, taking into account the villages, districts, and provinces mentioned in the sources. Other sources that mention the exact locations of sites were used to register their exact coordinates; examples include sites being used as museums and sites that have retained their original functions.

After the compilation of the initial inventory, fieldwork was organized to research the sites on the lists and gather information on their exact locations and current conditions. Three provinces so far have been studied thoroughly: Kayseri, Sivas, and Adana. Fieldwork resulted in the gathering of information and the completion of the

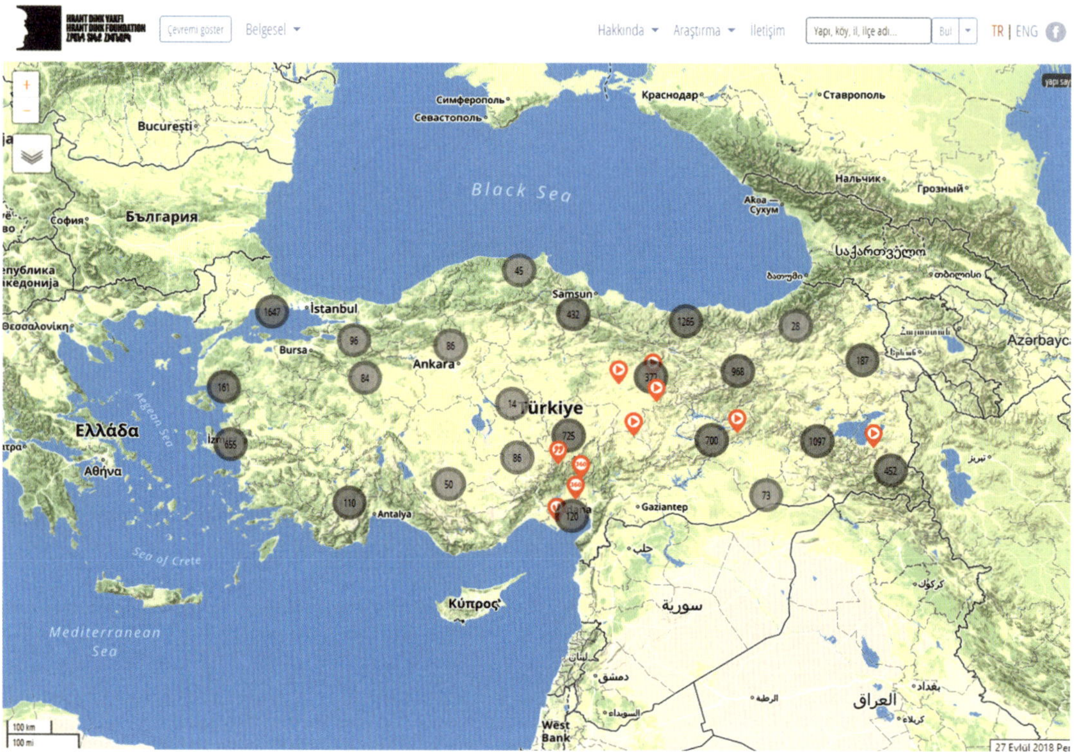

Fig. 1. Screenshot of the map interface after the 2018 update (Courtesy of the Hrant Dink Foundation Archives).

inventories of these provinces. The online map was updated accordingly, with visual data and exact locations added. Additional details about fieldwork in each province are provided below. General findings show that almost half the sites registered in the inventory cannot be located with current methods and capacity. The half that has been located, on the other hand, includes sites that cease to exist physically today but which were remembered by local communities we visited, attesting their locations. This makes the online map not only a visual representation of an inventory but also a document of witness testimonies.

The Hrant Dink Foundation plans on continuing this approach of mixing oral history, cultural heritage, and digital humanities research in its ongoing and future projects. Our experience shows that these disciplines can complement each other when considered in all the phases of the research—most importantly, in the planning phase. As explained below, oral histories deriving from short interviews are considered to be primary sources, essential informational components of the online map. In this regard, the Turkey Cultural Heritage Map is also an intrinsic part of the other humanities research projects carried out by the Hrant Dink Foundation, including conference series on Ottoman minorities in the turn of twentieth century, the oral history archive, and other cultural heritage projects. Our positive experience mixing sources and methods and our desire for a more accessible and holistic view of cultural heritage in Turkey led to a significant update of the online map in 2018, enabling it to accommodate more than just the Turkey Cultural Heritage Inventory. A new layer was added to host diverse types of media and visualize different kinds of findings, beyond mere geographic positions and spreadsheet data. Most importantly, this new update also enables an increased capability to present metadata and to share not only findings with users but also how research was conducted and the experiences of those working on the projects.

In the sections below I briefly present information on the Turkey Cultural Heritage Map, its inventory and methodology, its sources and research processes, and its associated fieldwork, as well as complementary workshops, programming, and the update implemented in 2018.

Inventory, Methodology, and Limitations

The online map currently contains approximately 9,500 sites that are registered in the inventory, 4,400 of which are Armenian, 4,100 Greek, 700 Syriac, and 300 Jewish. Information concerning Bulgarian, Georgian, and Catholic communities, as well as missionary institutions, was recorded in the course of the research but was not shared through the map, as the details were deemed insufficient.

"Ownership" and "belonging" are taken in their literal sense in the inventory. While aware of the limitations of such an approach and acknowledging exceptions, the patrons, owners, and users of a given building were assumed to belong to the same community. Buildings not owned or used by non-Muslim communities were omitted from the online map, regardless of the identity of their builders. From mosques to palaces, many examples can be found of architects or builders who belonged to minority communities; because such sites were not used by those communities themselves, they were considered not to be a part of the non-Muslim cultural heritage. The map intends in no way to impose decisive and final verdicts concerning the ownership and belonging of cultural heritage sites; nonetheless, such limitations were necessary to make the results of the research more meaningful.

For the purpose of defining what constitutes a site of cultural heritage belonging to a certain community, several operational rules were adopted, including chronological and typological limitations. For instance, buildings could be categorized as "Armenian" starting only from 301 CE, the year traditionally assumed to mark the Armenian state's adoption of Christianity as its official religion.

Greek cultural heritage was conceptualized in this study to comprise not only the Turkish Republican and Ottoman periods, but also the Byzantine period. "Byzantine" has been accepted here not as a separate ethno-religious category but as a period, and its beginning is set to 330 CE, the year traditionally accepted for the establishment of the capital of the Eastern Roman Empire in Constantinople and when, concurrently,

Christians were granted privileges. Public buildings from the Byzantine period, thus, were categorized as "Greek," unless they demonstrably belonged to another ethnic group.

Jewish cultural heritage in Turkey has been assumed to date as far back as the aftermath of the fall of Jerusalem in 70 CE, which corresponds to the (second) migration *en masse* of Jewish people westward. The identity of the users or owners of buildings, such as Romaniote, Sephardic, Karaite, or Ashkenazi, was specified when it was deemed important for the history of the buildings. The category of "Syriac" cultural heritage is taken as comprising the divisions of both the "Eastern and Western communities" and the "Eastern and Western Churches." The former relates to a geographical distinction starting from 37 CE, when the Syriac community is assumed to have adopted Christianity, and the latter corresponds to a theological schism between the years 430 and 460. While buildings belonging to these communities were all categorized as "Syriac," their ethnic and denominational sub-divisions (Aramaic, Jacobite, Syriac Catholic, Syriac Protestant, Chaldean, and Nestorian) were included in the detail sections of individual sites.

The inventory was limited by the site type—according to the original use—to include only synagogues, churches, chapels, monasteries, schools, orphanages, and hospitals. Sites that had more than one function, such as churches and schools or monasteries and orphanages, were recorded in the inventory as separate entries. Building subtypes, for example boys' or girls' schools, cathedrals or basilicas, were not recorded separately but, in most cases, were included in the details section. Many other types of cultural heritage were noted in the inventory but were not included in the online map; these include temples, castles, fountains, *hamams*, boarding houses, bridges, dams, irrigation and sewage systems, and mills. These and other types of entries provide examples of how widely the map or similar projects could be expanded.

The geographical boundaries of provinces, districts, and villages are generally prone to change over time. For inventory entries, the exact locations of many of which remain unknown, geographic attributions are approximate and were noted as such. Otherwise, administrative provinces and smaller components of them were adopted with current official names and borders, even though these have changed recently too.

The online inventory of settlement units in Turkey, Index Anatolicus (Nişanyan 2019), the dictionary of Ottoman place-names published by the General Directorate of the State Archives (Sezen 2017), and the five-volume Armenian-language masterpiece dictionary of place-names (Hagopyan, Melik-Pakhshyan, and Parseghyan 1986) were the main sources for matching old and current province and settlement/neighborhood/village names.

After 2004 legislation that required sixteen provinces to be redefined as metropolitan areas and 2013 legislation that increased the number of provinces designated as metropolitan areas from sixteen to twenty-nine, the villages of metropolitan areas were redefined as neighborhoods and the old "center" districts of these provinces-turned-metropolitan areas became obsolete, replaced by one or more newly named districts.

Main Sources

As mentioned above, the main sources of the Turkey Cultural Heritage Map can be divided into two main categories: first, the main lists that form the backbone of the inventory and, second, more local and specific sources. One of the main lists comes from seven bound notebooks containing church registers from the Turkish Presidency State Archives of the Republic of Turkey (the former Ottoman Archives of the Prime Minister's Office). Another main list is the bound notebook under Grand Vizierate classification containing an inventory of churches and schools put together by the state; it is undated but was likely produced in the early twentieth century. The notebook lists information on the provinces, districts, and villages where the buildings were located and, in some cases, information on the names and construction dates of the buildings, as well as the institutions to which they belonged. In addition, a set of bound notebooks, coded as A.DVNSKLS and under *Bab-ı Asafi* classification, includes documents on the construction, renovation, and repair of churches and schools from the 1870s to the mid-1910s. In addition to these notebooks, documents yielded by a catalog search with numerous keywords have been used, mostly dating from the nineteenth century. Other general sources were used to varying degrees of intensity, most importantly among them Vital Cuinet's (1891) late nineteenth-century survey of the Ottoman Empire.

The main source for Armenian cultural heritage is the list of Armenian churches and monasteries prepared between 1912 and 1913 by the Armenian Patriarchate of Istanbul for submission to the Ministry of Justice and Sects. This list was used in its Armenian translation (Safrastyan 1965). Other important sources on Armenian sites are the *hushamadyans*—memory books—published predominantly by compatriotic unions in the Armenian diaspora. Raymond Kévorkian and Paul Paboudjian's (1992) monumental work, *Les Armeniens*, was used in its Turkish version (Kévorkian and Paboudjian 2013). Among the main sources for Jewish cultural heritage is Naim Güleryüz and Kehribar İzzet's (2008) extensive survey, *Türkiye Sinagogları* [*Synagogues of Turkey*]. The main sources on Syriac buildings are Gabriel Akyüz's (1998) studies on the churches and monasteries around Mardin. Other than these main sources, *Agos* and *Shalom* newspapers' archives were used extensively.

Important non-textual sources for the online map are the oral history interviews that the Hrant Dink Foundation carried out for its *Sounds of Silence* publication series and oral history archive. The oral history methodology used to gather life stories was updated in 2017 to include questions specific to the use and experience of cultural heritage. Stories of churches and schools add a deeper dimension of human experience to the map, in addition to information on the approximate locations of sites. Further exploration of the incorporation of oral history in cultural heritage research is among the future directions the project will take.

The sources mentioned above constitute only the main sources for the online map, in addition to which a large number of books, articles, websites, testimonies, and unpublished documents in various languages are used to compile the inventory. Yet, the inventory is far from complete and is not intended to result in a static database. Ongoing research will expand it, and we also invite users to suggest changes and additions to the information on the map by filling out a form. This user feedback is an important information source that opens the door for contributions from a wide audience and renders participatory results much needed to keep the inventory up to date.

Fieldwork Activities

To complement the empirical information gathered through textual and oral sources, fieldwork was conducted between 2015 and 2017, focusing on the provinces of Kayseri, Adana, and Sivas. Other than the main sources used for the inventory, sources specific to these three provinces were consulted to provide more detailed information on the cultural heritage of these provinces. Through contact with local institutions and individuals, routes were planned in each of the provinces to visit villages and neighborhoods featured in the inventory in an effective manner. When a site, its remains, or its no-longer extant location was determined, its geographical coordinates were recorded by using a GPS device or GPS-enabled smartphone. If extant, the site was photographed. Observations on the access, current use, and physical and social states of the buildings were recorded, in addition to ownership information, when available. These data form the cultural heritage archive, according to which the data in the inventory and the online map are updated. The methodology included also conducting short interviews with local inhabitants, meaning some of the information gathered during the fieldwork was generated from the local community. This information was later compared with data coming from the other sources, and all compared results were recorded in the inventory.

The first fieldwork was conducted in Kayseri, in part because of its middle position between Armenian and Greek areas of central Anatolia. Many villages and towns of Kayseri had Muslim, Greek, and Armenian communities, which gives the province a unique setting. Two research trips were made to Kayseri in the summer and fall of 2015, resulting in on-site visits to 112 sites in the first trip and 69 sites in the second trip (Figs. 2 and 3). The Kayseri inventory comprising 344 entries was updated after the two field trips. The number of entries rose to a total of 387, of which 181 were visited. Of the visited sites, roughly half still bore extant evidence, while the remaining half were recoded based only on local memories.

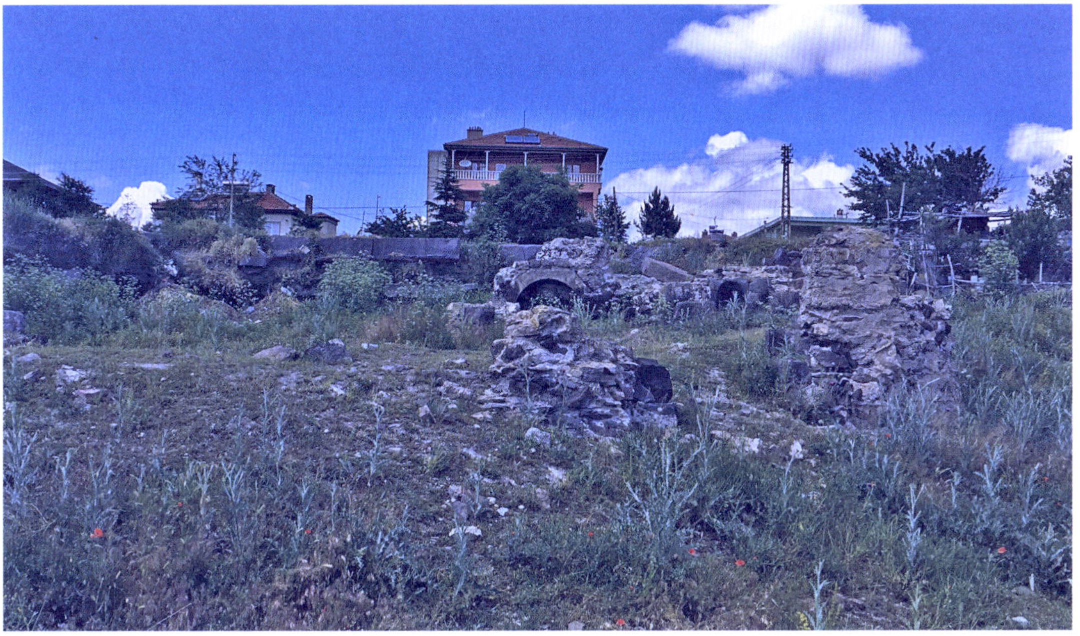

Fig. 2. Remains of Tomarza Surp Asdvadzadzin Monastery (Tomarza, Kayseri), today known to locals as "*manastır*" (Courtesy of the Hrant Dink Foundation Archives).

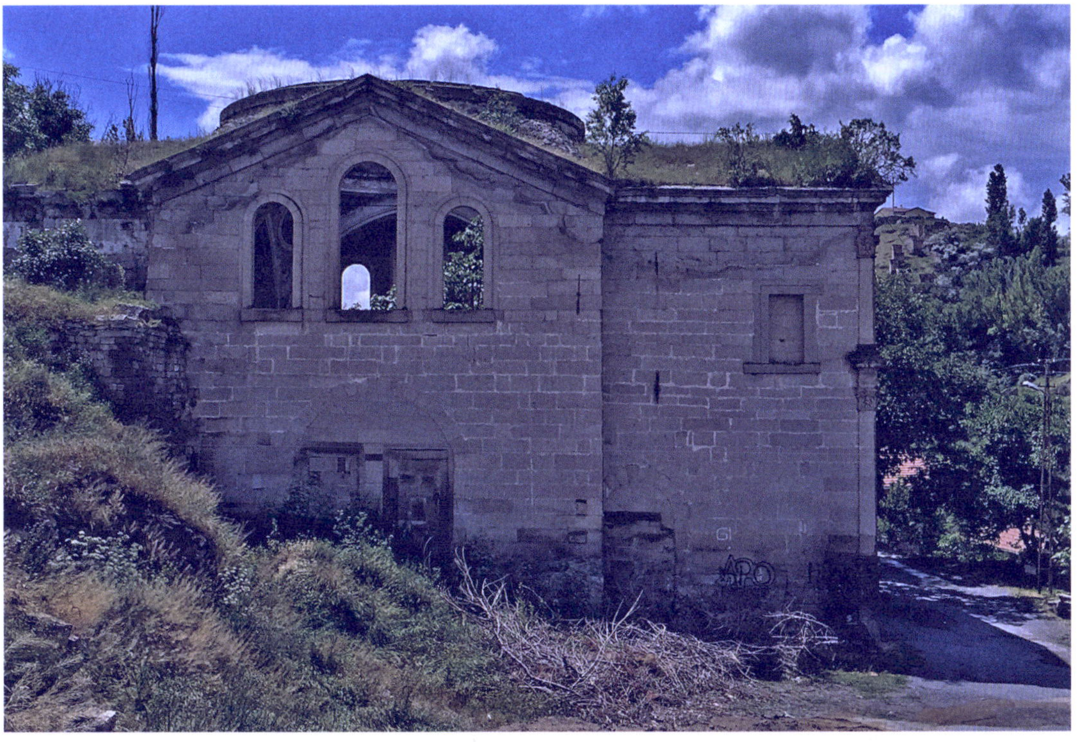

Fig. 3. The Surp Stepanos Church of Efkere (Melikgazi, Kayseri) (Courtesy of the Hrant Dink Foundation Archives).

The second province of fieldwork was Adana. Concerning Late Ottoman cultural heritage, three *hushamadyans* were used for the Armenian inventory, each relating to areas found today within the bounds of the modern province of Adana: Puzant Yeghiayan's (2012) work on the history of Adana; Misak Keleshian's (1949) work on the history of the Kozan-Sis region; and Hagop Boghosian's (1942) work on the history of Saimbeyli-Hadjin. Yeghyayan's work includes a list of property owned by Armenians in Cilicia under the French Mandate. Based on these works, we could say that the area included 33 Armenian sites in Adana center, 45 in Saimbeyli-Hadjin, and 47 in Kozan-Sis.

The Turkey Cultural Heritage Inventory had previously listed 80 sites in Adana. After adding these sites from Adana-specific sources, 62 of which were previously unattested, the number grew to 142. The two field trips in September 2016 and April 2017 added 17 sites to the Adana inventory for a total of 159 registered sites. During the field trips, 60 sites were visited on-site and documented (Fig. 4), while 85 sites could not be identified in the locations that were visited. This means that 38% of the sites were identified and their locations registered. Of the 85 not identified sites, 77 were located in urban locations in Adana center, Kozan-Sis, Saimbeyli-Hadjen, and Feke-Vahka. One result of the research in Adana shows that cultural heritage is more prone to disappearance in urban centers.

Fieldwork in Sivas was carried out in November 2016 and September-October 2017. With the new sites added to the list during the fieldwork, the Sivas inventory was increased from 335 to 426 sites (Figs. 5 and 6). Roughly two-thirds of all the sites were identified in 130 villages spread across nearly all its districts. Of the total 400 Armenian sites, 242 were identified; of the 21 Greek sites, 12 were identified. In addition to the Armenian and Greek sites, five sites recorded under the "missionary" category were all identified. The five "missionary" sites included three operated by the local stations of the Protestant missions, as well as two Jesuit Catholic sites. These five sites are currently not open for online access, but were included in the published book (Keshishian, Löker, and Polatel 2018c).

To treat the Seljukian Province of Rum, the Little Armenia of Byzantium, and the Ottoman *Vilayet* of Sivas as one unit not only does injustice to these entities but is also incorrect. To be able to observe the cultural diversification of Sivas, one needs to travel from Suşehri to Şarkışla, from Hafik to Gürün, and from Yıldızeli to Zara. The vastness of the province and its differing local histories posed challenges to the inventory work.

The existence of different layers of cultural heritage in the city of Sivas pertains equally to the whole province, with even more complexity stemming from regional and local influences. Sivas, Sevast, Siwas, Sebaste and other variations of the name itself derive from this layered history. Because of this complexity, it is important to differentiate between cultural heritage sites created or reused in at least three different periods: early medieval, tenth-twelfth centuries, and nineteenth century. These three periods represent, respectively, the early Christian era of the city, the time when the Byzantine

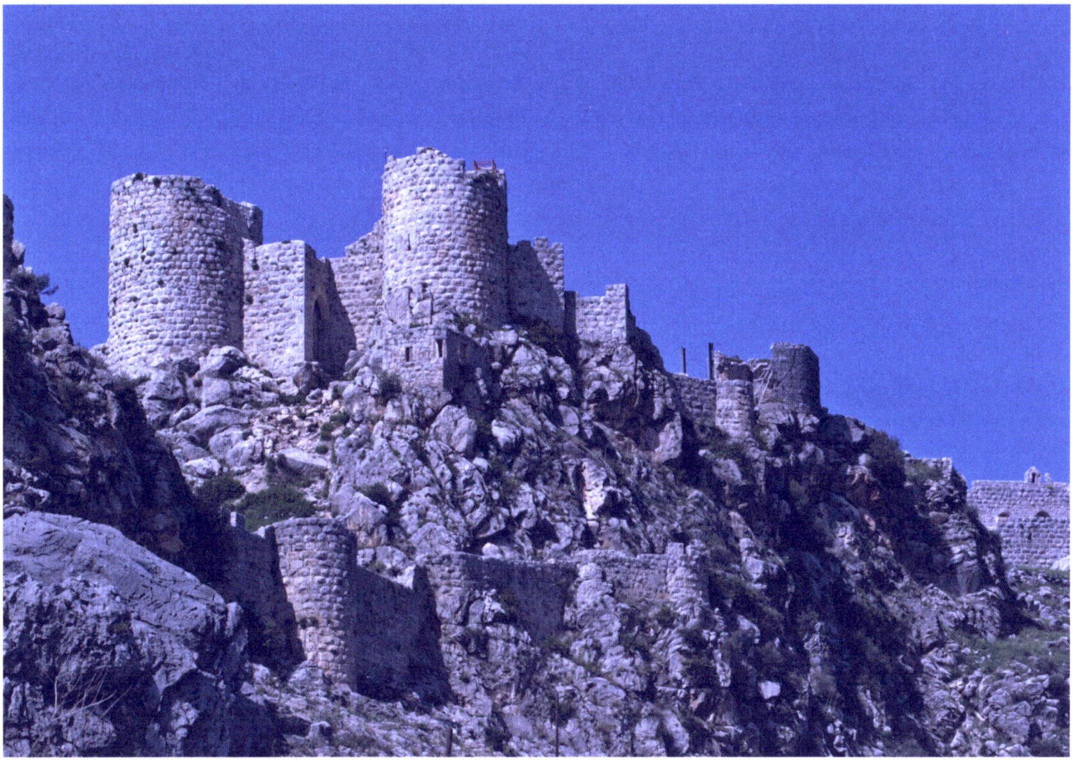

Fig. 4. Yılankale/Levon's Fortress (Levonkla) (Ceyhan, Adana), where the ruined chapel can be spotted (Courtesy of the Hrant Dink Foundation Archives).

Empire transferred Armenians living in the outskirts of the empire to central areas based on an agreement with the Armenian king Senekerim, and, finally, the industrial development, the Tanzimat, and the population boom of the nineteenth century. This diversification brings a multiplicity of sources, which in turn brings discrepancies in place-names. Different usages of place-names in Armenian and Ottoman sources, as well as the repetitive renaming of villages and settlements during the Republican era, pose challenges for the compilation of the inventory.

More general conclusions resulting from fieldwork include that the characteristics of village populations are decisive factors in the preservation of memories of cultural heritage. Many village inhabitants arrived in the area only after the departure of previous inhabitants, resulting in breaks in the chain of memory and challenging efforts to gather sufficient information for the inventory. The resettlements in Sivas of populations from the Balkans and the Caucasus have this effect to a great extent, while resettlements from other regions within Turkey, especially from the east and southeast, result in partial losses of memory. On the other hand, the most vivid memories were preserved in villages that once had mixed populations and whose current inhabitants remember ancestors who lived side-by-side with previous villagers as neighbors. Such memories

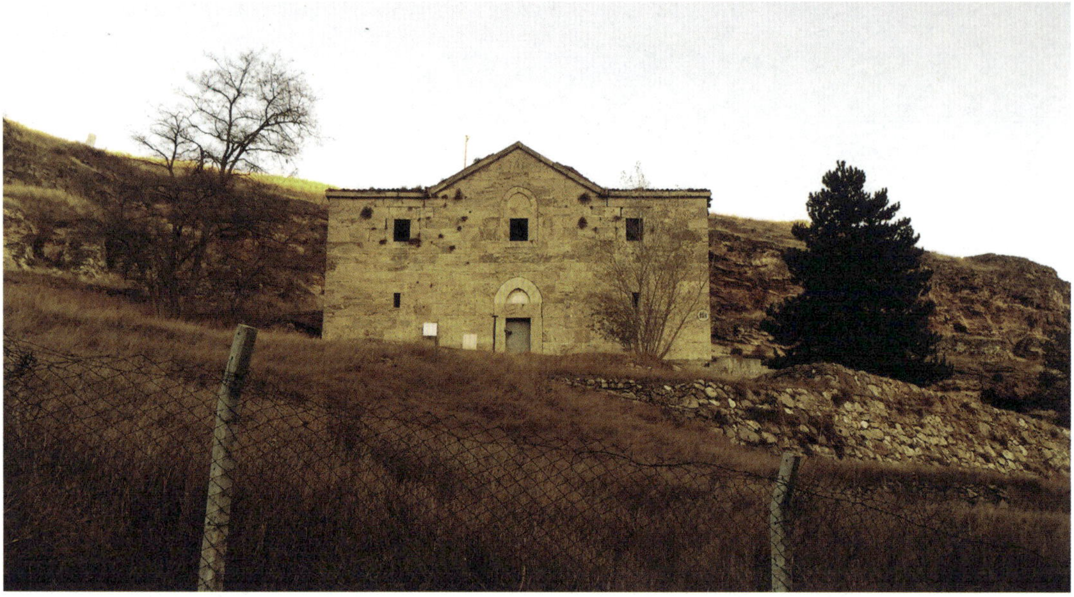

Fig. 5. Tavra Surp Sarkis Church, inside the military zone (Sivas Merkez, Sivas) (Courtesy of the Hrant Dink Foundation Archives).

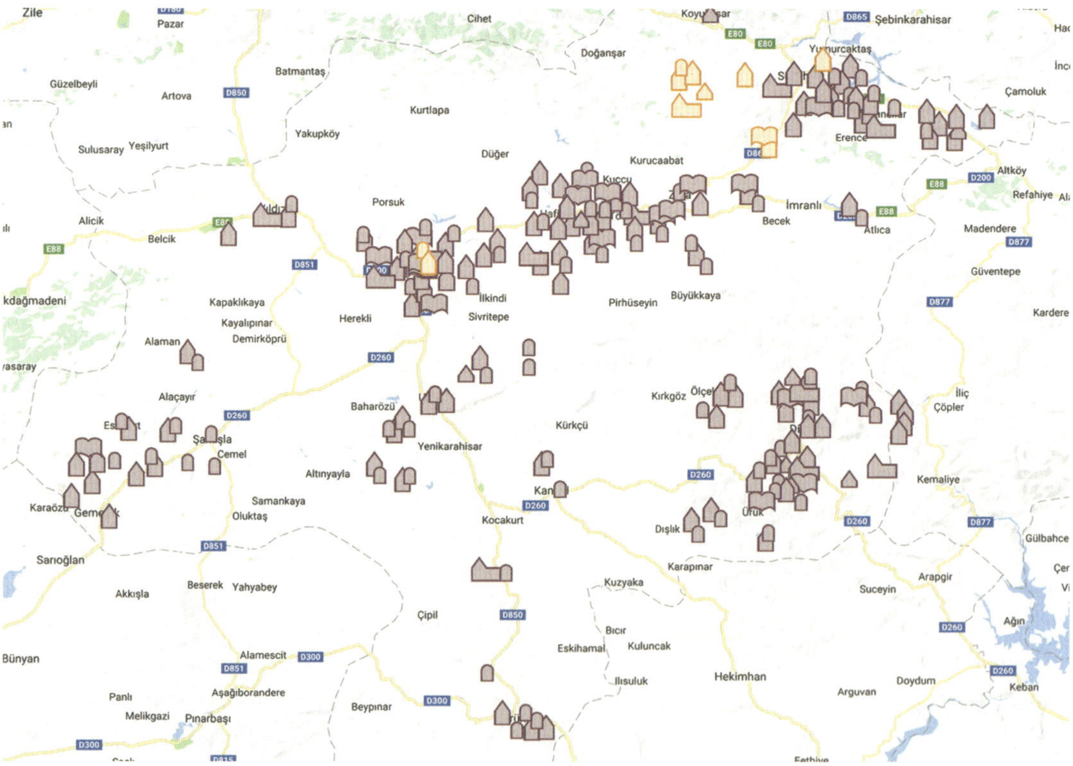

Fig. 6. A visualization on Google's "My Maps" function of all the GPS data gathered during the Sivas fieldwork (Courtesy of the Hrant Dink Foundation Archives).

were even more advantageous when non-Muslim community members, even if only few in number, continued to live in areas until more recently and passed on memories to current inhabitants. In addition to the mere presence of memories, the willingness to share them is also an important factor. Property ownership and other concerns keep people from sharing memories about the locations or stories of cultural heritage. In a few cases, the sites listed in the inventory could not be located because of such lack of willingness to share memories. In several cases, sources expressed their unwillingness to share information because of fear of restitution demands or legal actions that might endanger the ownership of their or their neighbors' properties.

A final more general conclusion concerns treasure hunting as a major factor of cultural heritage destruction in the region. Both locals and outsiders practice treasure hunting regularly despite measures taken by security authorities and the registration of sites for official protections. Registration is an important step toward the criminalization of destructive activities, yet mere registration falls short of having any tangible effect on the destruction process. Economic factors are cited by local sources as a major reason that treasure hunters pursue valuables in cultural heritage sites. This said, it should be noted that practitioners of treasure hunting have a grasp of cultural heritage sites and possess extensive knowledge of the landscape and geography of the area; in several cases, information provided by them during fieldwork found its way into the inventory.

Workshops, Participatory Research, and Parallel Publications

With participatory knowledge creation in mind, several workshops were organized as part of the fieldwork and in parallel to the research process. Oral history workshops were organized with the participation of students and locals in order to contribute to the oral history archive, which, in their turn, constituted an important component of the sources of the inventory as mentioned above. Oral history workshops were organized in Istanbul, Kayseri, Adana, Sivas center, and Sivas's Divriği district.

City tour workshops were also organized in Adana, Sivas, and Kayseri's Develi district. These workshops aimed to collect information on the multicultural heritage of these places. A "Bicycle Tour of Develi Fountains," organized with the support of local organizations, aimed to leverage the existing information in the online map to gather more information for it. A research workshop was held in Develi about the crafts of the city, considered intangible cultural heritage. In particular, shoemakers, dyers, glass smiths, masons, and bakers participated.

The audiovisual material collected during these workshops constitute the bulk of the material that can be viewed on the online map. The workshops proved to be an effective way of introducing the map to different audiences and helped to establish a two-way dialogue between the map and its users. Future directions and developments will include planning more participatory events and workshops in varying formats and with varying goals.

An integral part of the online map's research process has been the parallel publication of research reports concentrating on the fieldwork mentioned above. The first volume of the series was *Kayseri with Its Armenian and Greek Cultural Heritage* (Yılmaz 2016). The volume presented the results of the fieldwork in Kayseri jointly undertaken by the Hrant Dink Foundation and the Association for the Protection of Cultural Heritage (KMKD) and shed light on Kayseri's rich sociocultural past, tracing the physical and built environments left from the Armenian and Greek communities of the region. Additionally, it presented the relationship of the urban culture of Ottoman Kayseri to its rural hinterland by way of statistical information and visual material mainly derived from the map. The volume also included introductory articles that contextualized and shed light on the social and economic history of the Armenian and Greek communities of Kayseri.

Three more volumes were published in 2018, again summarizing the results of fieldwork undertaken between 2016 and 2017. The first volume (Keshishian, Löker, and Polatel 2018b) concentrated on the Develi district of Kayseri, which includes in its modern borders the old Armenian Everek and Fenese and the Greek Aygösten villages. Develi is presented as one of the rare cities of Anatolia in which Armenians had lived as a community in the Republican era. The community, in addition to its churches and schools, had a long-established history of economic development but then lost everything in 1915; afterward it succeeded in partially rebuilding and managed to maintain its identity as an Armenian community. With the help of the information gathered from the oral history interviews, the volume features a chapter which sheds light on the history of the Develi Armenians during the Republican era and also on the story of the last Armenian house that still stands today. The volume also features the translation of the memoir of one of the Develi Armenians, Hagop Arzumayan, penned in 1935, and also the translation of an article written by Jack Der Sarkissian, recounting the transformations that took place within the Develi Armenian community during the late nineteenth and early twentieth centuries.

The Adana volume (Keshishian, Löker, and Polatel 2018a) starts with a local chronology and continues with cultural heritage, featuring an inventory that consists of the information about 163 sites from the Adana inventory, also visualized on the online map. Presenting information on cultural heritage structures over a long time span, the book focuses on transformations that took place from Byzantine Cilicia, to the nineteenth century Çukurova Valley, to Republican Adana, and includes a memory tour that traces the multicultural period of the Adana city center. The book also includes a section that discusses the possibilities of reuse by addressing the Nuri Has Passage, the former Apkaryan School, and the relation of cultural heritage and "indigenous production landscapes," or areas where "ancient forms of practices for sustaining land and waterscapes" can be observed (Fig. 7) (Eken and Mevlat 2018). The correlation between nature and cultural heritage was explored with the contribution of the Doğa

Derneği [Nature Association], whose experts joined the Hrant Dink Foundation's fieldwork in the Aladağ district of Adana and the Divriği district of Sivas.

Like the Adana volume, the Sivas volume (Keshishian, Löker, and Polatel 2018c) starts with a local chronology and continues with cultural heritage, featuring an inventory that consists of the information about 430 structures, again also visualized on the online map. The volume gives information about cultural heritage structures of great temporal, geographic, and structural diversity. On the one hand, the book shows Sivas as a whole; on the other hand, it addresses the particularities of each region. It also features a memory tour of the Sivas city center. One of the chapters is dedicated to the city's modernization period, beginning in the mid-nineteenth century, and gives information about the Armenian schools of the city. The last chapter discusses possibilities for the reuse of the Tavra Surp Sarkis Church, located near the center of Sivas. Moreover, with the contributions of experts from the Doğa Derneği, the last chapter addresses the relationship of the "indigenous production landscapes" and the cultural heritage of Divriği.

Each of these books was accompanied by a stand-alone printed foldable map of the city center, which includes cross references with the online map. These publications help reach wider audiences, especially those living in the countryside, and also give the online map a published print presence. Next to the maps and the volumes, the

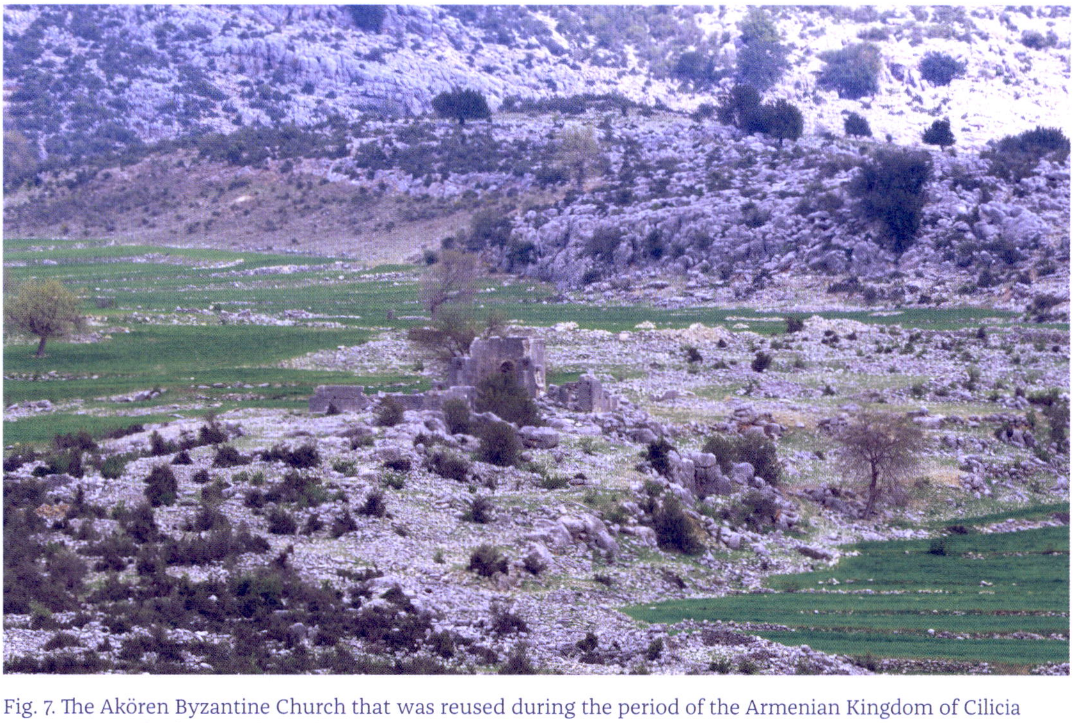

Fig. 7. The Akören Byzantine Church that was reused during the period of the Armenian Kingdom of Cilicia as a monastic facility (Aladağ, Adana). Here it is pictured surrounded by indigenous production landscapes (Courtesy of the Hrant Dink Foundation Archives).

online map is also accompanied by online publications on the Hrant Dink Foundation website. Case studies, further elaborations, and updates on the research process are periodically published on the website. These publications and the online map are not meant to duplicate one another but rather, since they result from the same research, offer complementary information and functionality.

Programming and the 2018 Update

An update to the online map in mid-2018 concentrated on expanding the layers of mapping not only to include geolocations and limited information on sites but also to include stories, experiences, and memories. New types of media were used to present this information, with topics presented in the form of stories and city tours, accompanied by 360-degree panoramas, videos, and audio and image files. Going beyond tangible cultural heritage, the online map now also contains findings on intangible cultural heritage. Believing that new media forms can provide effective tools for the revival of memory, the infrastructure of the website was redesigned and the code completely rewritten to accommodate new types of information. The first version of the web application was based on Google Maps and used several JavaScript libraries for queries. The new version uses Mapbox maps and Leaflet for queries. Leaflet is an open-source JavaScript library for mobile-friendly interactive maps. This shift, along with several edits in the search and browsing functions, was deemed necessary first because of changes in the Google Maps API that led the old version to suffer from malfunctions and, second, because we aimed to present the online map through a more versatile, mobile-friendly, and responsive portal.

Conclusion

The Turkey Cultural Heritage Map is an evolving tool for researchers and enthusiasts, for those who want to learn about the communities represented on the map, and for those who want to look for cultural heritage sites in their own villages. One of the aims of the project was to produce a map that visualizes not only what is on the ground today but also what was once present but no longer survives. In that regard, the numbered clusters and the icons all show not statistics, but life, culture, and cultural heritage and the communities built up around them.

Works Cited

(A full list of works cited in the database is available under the "Bibliography" tab at https://turkiyekulturvarliklari.hrantdink.org/#content-view/4)

Akyüz, Gabriyel. 1998a. *Deyrulzafaran Manastırı'nın Tarihi*. Mardin: Kırklar Kilisesi Yayınevi.

———. 1998b. *Mardin* İli'nin *Merkezinde, Civar Köylerinde ve* İlçelerinde *Bulunan Kiliselerin ve Manastırların Tarihi*. Istanbul: Resim Matbaacılık.

Boghosyan, Hagop. 1942. *Հաճընի ընդհանուր պատմութիւնը եւ շրջակայ Գօզան-տաղի հայ գիւղերը* [General History of Hadjin and the Armenian Villages of the Nearby Kozan Mountain]. Los Angeles: Hadjin Compatriotic Reconstruction Union.

Cuinet, Vital. 1891. *La Turquie d'Asie géographie administrative*, vol. 2. Paris: Ernest Leroux.

Eken, Güven and Alen Mevlat. 2018. "Uncovering a Neglected Heritage: Indigenous Production Landscapes of Divriği." In *Sivas with its Armenian Cultural Heritage*, edited by Vahakn Keshishian, Koray Löker, and Mehmet Polatel, 135-43. Istanbul: Hrant Dink Vakfı Yayınları.

Güleryüz, Naim, and Kehribar İzzet. 2008. *Türkiye Sinagogları* [Synagogues of Turkey], 2 vols. Istanbul: Gözlem Gazetecilik Basın ve Yayın.

Hagopyan, Tateos K., Sdepan D. Melik-Pakhshyan, and Hovannes K. Parseghyan. 1986-2001. *Հայաստանի և հարակից շրջանների տեղանունների բառարան* [Dictionary of Place-Names of Armenia and Neighbouring Areas], 5 vols. Yerevan: Yerevan University Press.

Hrant Dink Foundation. n.d. Cultural Heritage Map (website). Accessed May 20, 2019. www.turkeyculturalheritage.com.

Keleshian, Misak. 1949. *Սիս-մատեան: Պատմական, բանասիրական, տեղեկագրական, ազգագրական եւ յարակից պարագաներ* [Book of Sis: History, Philology, Geography, Ethnography and the Surrounding Areas]. Beirut: Hay Djemaran Publishing House.

Keshishian, Vahakn, Koray Löker, and Mehmet Polatel, eds. 2018a. *Adana with its Armenian Cultural Heritage*. Istanbul: Hrant Dink Vakfı Yayınları.

https://hrantdink.org/en/hdv-publications/59-cultural-heritage/1404-adana-with-its-armenian-cultural-heritage.

———, eds. 2018b. *Develi with its Armenian Cultural Heritage*. Istanbul: Hrant Dink Vakfı Yayınları.

https://hrantdink.org/en/hdv-publications/59-cultural-heritage/1403-develi-with-its-armenian-cultural-heritage.

———, eds. 2018c. *Sivas with its Armenian Cultural Heritage*. İstanbul: Hrant Dink Vakfı Yayınları.

https://hrantdink.org/en/hdv-publications/59-cultural-heritage/1405-sivas-with-its-armenian-cultural-heritage.

Kévorkian, Raymond H., and Paul B. Paboudjian. 1992. *Les Arméniens dans l'Empire ottoman à la veille du génocide*. Paris: Editions d'art et d'histoire ARHIS.

———. 2013. *1915 Öncesinde Osmanlı İmparatorluğu'nda Ermeniler*. Translated by Mayda Saris. Istanbul: Aras Yayıncılık.

Nişanyan, Sevan. 2019. *Index Anatolicus Türkiye Yerleşim Birimleri Envanteri*. https://nisanyanmap.com/.

Safrastyan, Aram Khaçaduri. 1965. "Կոստանդնուպոլսի Հայոց Պատրիարքարանի Կողմից Թուրքիայի Արդարադատութեան եւ Դաւանանքների Մինիստրութեան Ներկայացված Հայկական Եկեղեցիների եւ Վանքերի Ցուցակներն ու Թակրիրները 1912–1913" [The Lists and *Takrirs* of the Armenian Churches and Monasteries Presented to the Ministry of Interior of Turkey by the Armenian Patriarchate of Constantinople, 1912–1913]. *Echmiadzin* 1: 41–47.

Sezen, Tahir, ed. 2017. *Osmanlı Yer Adları*. Ankara: T.C. Başbakanlık Devlet Arşivleri Genel Müdürlüğü.

Yeghiayan, Puzant. 2012. *Ատանայի Հայոց Պատմութիւն* [History of Adana Armenians]. Antilias: Holy See of Cilicia.

Yılmaz, Altuğ, ed. 2016. *Kayseri with its Armenian and Greek Cultural Heritage*. Istanbul: Hrant Dink Vakfı Yayınları.

https://hrantdink.org/en/hdv-publications/59-cultural-heritage/340-kayseri-with-its-armenian-and-greek-cultural-heritage.

Yılmaz, Serpil Sönmez. 2011. *Osmanlı Klasik Çağında Sivas*. Istanbul: Kitabevi.

Zartanyan, Yervant. 1972. *Պատմագիրք Տիվրիկ Քաղաքի* [History Book of the City of Divriği]. Beirut: Hamazkayini Vahe Setyan Dbaran.

CONTRIBUTORS

Ayşe Nur Akdal received her BA degree in 2012 from the Urban and Regional Planning Department, Mimar Sinan Fine Arts University, and went on to do an MA in the Atatürk Institute for Modern Turkish History at Boğaziçi University, where she is currently a PhD candidate. She has been a research fellow at the Istanbul Studies Center (ISC), Kadir Has University, since 2016 and her MA thesis on "Market Gardens and Gardeners of Ottoman Istanbul" was published in 2017.

<aysenur.akdal@gmail.com>

Elton Barker is a Reader in Classical Studies at The Open University, having previously taught at Oxford (Christ Church). His studies in Classical and Greek Civilization at the University of Leeds (BA and MA respectively) took him to Ohio State for a Masters in Greek and Latin, and thence to Pembroke College, Cambridge for his doctorate on the verbal agon, or contest of words, in Homer, historiography, and tragedy, supervised by Simon Goldhill and Paul Cartledge. Since 2008 he has been at the forefront of collaborative efforts to develop digital resources for exploring historical places, notably as the community director of Pelagios.

<elton.barker@open.ac.uk>

Anuja Dangol is a research associate working for the Sagalassos Archaeological Research Project since 2016, where she has been tirelessly developing the Sagalassos Integrated Information System (SIIS). In general, her work focuses on database modelling (UML) and design (SQLITe, PostgreSQL/PostGIS), and the creation of an application facilitating (geo) data collection—making use of Python and QGIS—for archaeological fieldwork. She is currently managing the project based on Semantic web technologies and Linked Open Data in Cognizone (https://www.cogni.zone/). Her accumulated professional experience includes GIS, Geo-ICT, Linked Open Data, Remote Sensing, and environmental statistical analysis.

<anuja.dangol@cogni.zone>

Nicholas de Lange is Professor of Hebrew and Jewish Studies Emeritus in the University of Cambridge. He specializes in the history of the Jewish minority in the Byzantine empire. His publications include *Greek Jewish Texts from the Cairo Genizah* (Tübingen: Mohr Siebeck, 1996); "Research on Byzantine Jewry: the state of the question," in *Jewish Studies at the Central European University IV, 2003-2005*, edited by András Kovács and Michael L. Miller (Budapest: CEU, 2006), 41-51; and "Byzantium," in *The Cambridge History of Judaism, Vol. 6. The Middle Ages: The Christian World*, edited by Robert Chazan (Cambridge University Press, 2018), 76-97.

<nrml1@cam.ac.uk>

Claudia Engel is Academic Technology Specialist and Lecturer at the Department of Anthropology at Stanford University and affiliate of the Center for Interdisciplinary Digital Research at the Stanford Libraries. She has been involved in numerous projects that focus on the innovative potential of new technologies for teaching and research. From 2012 to 2014 she served as database developer for the Çatalhöyük Research Project.

<cengel@stanford.edu>

Piet Gerrits is a Research Fellow working for the UrbanOccupationsOETR project at Koç University in Istanbul. He received his undergraduate degree in Archaeology with a specialization in GIS at the Vrije Universiteit, Amsterdam. Working with the Barcın Höyük excavation and UrbanOccupationsOETR projects during his graduate studies at Koç University (MA 2018), he gained an interest in developing spatial databases and conducting computational research in archaeology and history. His current research focuses on the development of transport, historical maps, and Turkish population censuses.

<pgerrits15@ku.edu.tr>

H. Murat Güvenç received his PhD from Middle East Technical University, Ankara. His academic interests concentrate on the Visualization of Spatial Data, Industrial Geography, and Urban History. He co-authored the institutional history of the Real Estate Bank of Turkey (*Emlak Bankası 1926–1998*) and the *Electoral Atlas of Turkey 1950–2009*. He was on the curatorial team of the *Istanbul 1910–2010* exposition. Since 2014 he is associated with Kadir Has University and directing its Istanbul Studies Center. Professor Güvenç is Vice President of the Social and Economic Research Foundation of Turkey (TESEV) and directs a TUBITAK project on the emergence of Istanbul City Region.

<hmguvenc@gmail.com>

Piraye Hacıgüzeller is an archaeologist (MA, PhD) and civil engineer (BSc) who has worked for the Sagalassos Project on and off since 2005. Her research interests concern topics in digital and spatial archaeology on which she has published extensively, including co-edited books such as *Re-Mapping Archaeology* (Routledge, 2019) and *Archaeological Spatial Analysis* (Routledge, 2020). She is currently a post-doctoral researcher at the Department of Archaeology of Ghent University and the Ghent Center for Digital Humanities. She serves as an assistant director of the Sagalassos Archaeological Research Project, where she designs, coordinates, and advises on various digital workflows. She is also the co-chair of the Geohumanities Working Group within DARIAH-EU, a pan-European digital infrastructure for arts and humanities scholars.

<Piraye.Haciguzeller@UGent.be>

M. Erdem Kabadayı teaches in the Department of History, Koç University. Until 2005 he has worked mainly on the economic, financial, and labor history of the Ottoman Empire. Within the UrbanOccupationsOETR project, he is pursuing his academic career further as an economic historian focusing on the Ottoman Empire, Bulgaria, Greece, and Turkey.

<mkabadayi@ku.edu.tr>

Vahakn Keshishian is a graduate of Haigazian University and American University of Beirut with degrees in Political Science and Public Administration. Between 2015 and 2018 he was Research Coordinator and History Program Coordinator at the Hrant Dink Foundation in Istanbul. He has participated in the translation, authorship, and editing of many of the Hrant Dink Foundation publications.

<vahaknkeshishian@hrantdink.org>

Daniel Knitter is postdoctoral researcher in the Physical Geography unit of Christian-Albrechts-Universität zu Kiel. He is a geographer and works with prehistoric and classical archaeologists in different research projects in central Europe, Greece, and Turkey. His research focuses on questions of human-society-landscape-environment interactions. His main methodological tools are quantitative spatial analyses using models as heuristic devices. Daniel is interested in critical physical geography, reproducible research, philosophy of science, and inter- and transdisciplinarity.

<knitter@geographie.uni-kiel.de>

Turgay Koçak is a PhD student at Istanbul 29 Mayıs University. After obtaining his BA in History from Istanbul University, he received his MA in History from Istanbul 29 Mayıs University in 2016 with a thesis on "Battles and Operational Area in the Ottoman-Safavid Struggle 1578–1590." His current research interests include Ottoman Military History, Ottoman Balkans History, and Digital Humanities approaches to History.

<tukocak@ku.edu.tr>

Bernhard Ludwig is a PhD student in Geography at Freie Universität Berlin and works in the Istanbul Department of the German Archaeological Institute (DAI) on the project "Transformation of the Pergamon Micro-Region between the Hellenistic Period and the Roman Imperial Age." Having studied both field archaeology and landscape archaeology, his research focuses on questions of human-environment-interdependencies with a methodological focus on spatial analysis.

<Bernhard.Ludwig@dainst.de>

Dominik Lukas (MA 2008, University of Leipzig) is the Developer of the Living Archive Project and Database Developer of the Çatalhöyük Research Project at Stanford University. He studied prehistoric archaeology and philosophy (University of Leipzig, Université-Marc-Bloch/Strasbourg). In 2008 he completed his master's thesis on the archaeological concept of the late iron age "Oppidum." From 2008 to 2012 he worked at the German Archaeological Institute (DAI) and then at the Excellence Cluster Topoi in Berlin as coordinator for research databases.

<dlukas@stanford.edu>

Camilla Mazzucato is a PhD Candidate in Anthropology at Stanford University. She has been part of the Çatalhöyük Research Project team since 2009 as the main GIS specialist. She completed an MA in Archaeology at Bologna University and MSc in GIS and Spatial Analysis at UCL, and she worked (2011-2012) as a pre-doctoral researcher at the Institute of Social and Cultural Anthropology of Oxford University. Most recently, Camilla held a PhD fellowship in Istanbul at ANAMED (Koç University Research Center for Anatolian Civilizations).

<camimnazz@stanford.edu>

Osman Özkan is a PhD student at Mimar Sinan Fine Arts University. He received his MA in history from Istanbul 29 Mayıs University in 2016 with a thesis focusing on the "Ottoman Naval Power in Early Modern Age." He has worked on projects on Ottoman labor history, environmental history, and urbanism in the Ottoman Empire and is interested in Ottoman maritime history, history of transportation networks, and digital humanities.

<osozkan@ku.edu.tr>

Jeroen Poblome is a Professor of Classical Archaeology at the Department of Archaeology, Art History, and Musicology at the University of Leuven, where he received his PhD in Archaeology. His extensive archaeological fieldwork in Belgium, Greece, and Turkey led to his directorship of the ICRATES project and the Sagalassos Archaeological Research Project. Jeroen actively publishes and edits as well as engages in public outreach with specific research expertise on ancient economies and the long-term development of social-ecological systems at regional scales. He is member of numerous professional bodies, boards, associations, institutes, and networks that promote the field of classical and especially Roman archaeology.

<jeroen.poblome@kuleuven.be>

Michael Połczyński is an Assistant Teaching Professor at Georgetown University in Washington DC. His current projects, along with the OpenOttoman Gazetteer, include a book on Polish-Lithuanian Muslims in the court of Ottoman Sultan Suleyman I and a book on English travelers in Early Modern Poland.

<Michael.Polczynski@georgetown.edu>

Christopher H. Roosevelt (PhD 2003, Cornell) is a Professor in Koç University's Department of Archaeology and History of Art, Director of the Research Center for Anatolian Civilizations and the Kaymakçı Archaeological Project, and Co-Director of Gygaia Projects. His work on the archaeology of Bronze and Iron Age Anatolia has been published with Cambridge University Press (*The Archaeology of Lydia from Gyges to Alexander*, 2009), Koç University Press (*Gyges'ten Büyük* İskender'e *Lydia Arkeolojisi*, 2017), and in numerous archaeological journals. He remains a member of the Archaeological Institute of America and the British Institute in Ankara, a corresponding member of both the German Archaeological Institute and the Turkish Institute of Archaeology, and a fellow of the Society of Antiquaries, London.

<chroosevelt@ku.edu.tr>

Amy Singer (PhD 1989, Princeton) is the Hassenfeld Professor of Islamic Studies in the Department of History at Brandeis University and professor emerita from the Department of Middle Eastern and African History at Tel Aviv University. Her research began with an in-depth study of the relations between Ottoman officials and Palestinian peasants (*Palestinian Peasants and Ottoman Officials*, 1994), leading to a study of the Haseki Sultan public kitchen (*imaret*), endowed in mid-sixteenth-century Jerusalem (*Constructing Ottoman Beneficence*, 2002), and later to a broader study of benevolent giving (*Charity in Islamic Societies*, 2008), each monograph appearing also in Turkish translations. Her current research explores how the city of Edirne participated in the formation of the Ottoman state and society in the first half of the fifteenth century.

<amysinger@brandeis.edu>

INDEX

E

F

G

H

N

O

Ö

P

Q

R

S